EUROPEAN RESEARCH ON VARROATOSIS CONTROL

T0262957

The EC Experts' Group Meeting on 'European Research on Varroatosis Control' was organized by:

Commission of the European Communities
Directorate-General Agriculture

in collaboration with

Institut für Bienenkunde (Polytechnische Gesellschaft)
an der Universität Frankfurt am Main

Proceedings of a Meeting of the EC Experts' Group / Bad Homburg 15-17 October 1986

EUROPEAN RESEARCH ON VARROATOSIS CONTROL

Edited by
R.CAVALLORO
Commission of the European Communities, Joint Research Centre, Ispra

Published for the Commission of the European Communities by
A.A.BALKEMA / ROTTERDAM / BROOKFIELD / 1988

The texts of the various papers in this volume were set individually by typists under the supervision of each of the authors concerned.

CIP-DATA KONINKLIJKE BIBLIOTHEEK, DEN HAAG

European

European research on varroatosis control: proceedings of a meeting of the EC experts' group, Bad Homburg, 15-17 October 1986 / ed. by R.Cavalloro. – Rotterdam [etc.]: Balkema. – Ill.
Publ. for the Commission of the European Communities. – With index
ISBN 90 6191 846 4 bound
SISO 597.81 UDC 591.2:595.799
Subject heading: Varroatosis.

Authorization to photocopy items for internal or personal use, or the internal or personal use of specific clients, is granted by A.A.Balkema, Rotterdam, provided that the base fee of US$1.00 per copy, plus US$0.10 per page is paid directly to Copyright Clearance Center, 27 Congress Street, Salem, MA 01970. For those organizations that have been granted a photocopy license by CCC, a separate system of payment has been arranged. The fee code for users of the Transactional Reporting Service is: 90 6191 846 4/88 US$1.00 + US$0.10.

Publication arrangements: *P.P.Rotondó,* Commission of the European Communities, Directorate-General Telecommunications, Information Industries and Innovation, Luxembourg

LEGAL NOTICE: Neither the Commission of the European Communities nor any person acting on behalf of the Commission is responsible for the use which might be made of the following information.

EUR 10845 En

Published by
A.A.Balkema, P.O.Box 1675, 3000 BR Rotterdam, Netherlands
A.A.Balkema Publishers, Old Post Road, Brookfield, VT 05036, USA

ISBN 90 6191 846 4

© 1988 A.A.Balkema, Rotterdam
Printed in the Netherlands

Foreword

Varroa jacobsoni Oud. is continuing to spread in Europe and is invading new territories, with serious economic repercussions. Bee-keeping and, more generally, agriculture, are damaged in various ways, as the expansion of this very harmful mite has direct effects on the productions of the bee itself and on the fundamental and irreplaceable contribution of the bee to the cross-fertilization of agricultural crops.

The situation has stimulated the Commission of the European Communities which, as well as setting up concrete initiatives to help the sector affected, is developing actions to sustain bee-keeping and organizing scientific research to gain greater knowledge of the bio-ethological characteristics of the mite to impede the expansion of the infestations and to find a correct control method against the Varroa-mite.

Bee-keeping has an enormous economic value in the EC-Member countries which include about five million bee colonies within their boundaries. Thus the phenomenon of the persistence of the continuous propagation of this very damaging mite in Europe is closely followed by the Community Authorities, who act through various short and mean term actions.

This includes important specific research on the Varroa-mite which is being carried out as part of a five-year "Integrated Plant Protection" programme (1984-1988). This is being carried out jointly by the EC-Member countries in the framework of common and coordinated activities by the CEC, Directorate-General Agriculture.

In this context an Experts' Group Meeting was organized in Bad Homburg, F.R. Germany, to collect together the European specialists in this sector to examine the present situation of Varroatosis, the progress made up to now in the study of the biology and behaviour of Varroa jacobsoni, of its relationships with bees, both adult and in the various stages of their life-span, of the effects of the mite attacks, and of the possibilities of defence taking into consideration economic, toxicological and ecological aspects.

The publication illustrates the course of the meeting, presents all the papers given "in extenso", shows an up-to-date picture of the present status of Varroatosis, and above all considers the advanced methods for detecting and controlling Varroa jacobsoni.

Fifty-three highly qualified experts from eleven European countries attended the meeting. They discussed a total of 40 scientific contributions.

The volume follows the development of the work-meeting, which were subdivided in working sessions, taking into account the current situation of Varroatosis in Europe, the biology and behaviour of Varroa-mite, the parasite-host relationship, the secondary infection micro-organisms transferred from the Varroa in the hive, as well as different diagnostic and therapeutic control methods.

An important clarification of the Varroatosis problem was achieved, with the presentation of a wide range of lines of activity under way which are at the basis of future developments of scientific and applied research.

The meeting was very successful, as can also be seen by the proceedings printed by the CEC, which will certainly form an important and valid reference point for European research on Varroatosis control.

R. Cavalloro

ORGANIZING COMMITTEE

Cavalloro Raffaele
Principal Scientific Officer CEC, Responsible of "Integrated Plant Protection" Programme, Joint Research Centre, I-Ispra

Koeniger Nikolaus
Head Institut für Bienenkunde (Polytechnische Gesellschaft) an der J.W. Goethe Universität Frankfurt am Main, D-Oberursel

STRUCTURE OF THE MEETING

Introduction
Opening address by R. Cavalloro and F. Schieferstein

Sessions

Session 1: Current situation of Varroatosis in European Countries   Chairman: R. Cavalloro

Round table on Varroatosis in Europe   Chairman: D.A. Griffiths

Session 2: Biology of Varroa jacobsoni Oud.   Chairmen: M.D. Ifantidis
J. Beetsma

Session 3: Microbes and laboratory techniques   Chairman: H. Hansen

Session 4: Control methods   Chairmen: R. Borneck
O. Van Laere
F. Frilli

Conclusions
General discussion, conclusions and recommendations by N. Koeniger

ORGANIZING SECRETARIAT

Miss Christl Rau
Institut für Bienenkunde
Karl-von Frisch-Weg, 2
D-6370 Oberursel

PROCEEDING DESK

Mr. Pier Paolo Rotondò
Commission of the European Communities
Directorate-General Telecommunications, Information, Industries & Innovation
L-2920 Luxembourg

Table of contents

Introduction

Session 1. *Current situation of Varroatosis in European countries*
Round table on Varroatosis in Europe

Session 2. *Biology of* Varroa jacobsoni *Oud.*

Session 3. *Microbes and laboratory techniques*

Session 4. *Control methods*

Conclusions

Introduction

Opening address – Research activities on Varroatosis promoted by the Commission of the European Communities

R.Cavalloro

Principal Scientific Officer CEC, Responsible of 'Integrated Plant Protection' Programme

It is my fortunate task to welcome all the participants to this Experts' Group Meeting on the problem of Varroatosis in European countries. I therefore do so, both in the name of the Commission of the European Communities, Directorate-General for Agriculture, and personally, in the hope that the work will be greatly fruitful.

The presence of a considerable number of eminent qualified experts from eleven European countries is in itself a guarantee of success, and this must stimulate us even further to an open exchange of ideas and to an objective and open discussion so that we can direct the control against the very damaging mite Varroa jacobsoni Oud. in the best way, on the basis of the results obtained up to now and on the positive prospects which can be predicted in the short term.

I should like to thank the Authorities of the Federal Republic of Germany most sincerely for the help offered to us in organizing this meeting: in particular Dr. H. Pittler, representative of the German Federal Ministry of Agriculture and Forestry of Bonn, and Dr. E. Schieferstein, President of the German Bee-keepers' Association, who honour us with their presence.

A special thanks from the heart to Prof. N. Koeniger, Head of the Institute for Bee-keeping of the University of Frankfurt, and to his collaborators in Oberursel, for their close collaboration, which ensure us a useful, constructive and pleasant meeting, in an excellent atmosphere of true cordiality.

After the meeting promoted by the Commission of the European Communities in Wageningen, Netherlands, in 1983, to discuss the situation and the needs of the problems which had recently emerged at the level of the EC-member countries for the progressive extension of V. jacobsoni in Europe, and at Thessaloniki, Greece, in 1984, reviewing the actions in progress and presenting the orientations for future common activities, this Bad Homburg meeting considers more specifically European research from the point of view of economical and healthy methods of controlling Varroatosis.

The continuing spread of the Varroa-mite, which is being followed carefully by Community Organizations at all levels, Commission, Council, and

Tab. I - EC-Institutes involved in common activities on Varroa-mite re-
       searches

Belgium
   . Rijkstation voor Nematologie en Entomologie - Merelbeke
   . Laboratoire de Bio-Ecologie - Bruxelles
   . Laboratoire d'Ecologie et de Bio-Geographie - Louvain-la-Neuve

Denmark
   . Statens Bisygdomsnaevn - Lyngby

Federal Republic of Germany
   . Institut fur Bienenkunde - Oberursel
   . Tierhygienisches Institut - Freiburg

France
   . Station de Recherches sur l'Abeille et les Insectes sociaux - Bures-
     sur-Yvette
   . Institut Technique de l'Apiculture - Bures-sur-Yvette
   . Laboratoire National de la Pathologie des Abeilles - Nice

Great Britain
   . Rothamsted Experimental Station - Harpenden
   . Slough Laboratory, ADAS - Slough

Greece
   . Agricultural Research Station of Halkidiki - Nea Maudania
   . Laboratory of Apiculture, School of Agriculture - Thessaloniki
   . Laboratory of Sericulture and Apiculture, College of Agricultural
     Sciences - Athens

Ireland
   . Agriculture House - Dublin

Italy
   . Istituto Sperimentale per la Zoologia Agraria - Cascine del Riccio
   . Istituto di Difesa delle Piante - Udine

Netherlands
   . Research Centre for Insect Pollination and Beekeeping 'Ambrosiushoeve'-
     Hilvarenbeek
   . Department of Entomology and Virology, Landbouwhogeschool - Wageningen

----

European Parliament, has already led to the development of concrete actions for helping the sector involved in various ways, considering that this pest of bees may affect the irreplaceable services of bees in the cross-fertilization of agricultural crops and in pollination programmes, as well as having an influence on bee productions and thus on bee-keeping itself.

The urgent setting up of a first one-year programme has led to close collaboraion between the Member countries of the European Communities, where research teams from different countries with experts from different fields have in profitable agreement begun valid cooperation on research themes which range from the biology and behaviour of the mite to the ways of diffusion, from the quick identification of Varroatosis in bee-hives and from the role of semio-chemicals in host detection to the appearance of secondary infection micro-organisms transferred by the Varroa-mite, from the standardization of techniques and working methods to research into efficient control with products which have no negative side effects on the bee and its products.

This one-year programme has enabled us to obtain, preliminary, more exact knowledge and thus to identify valid lines of research more correctly.

One should stress the active collaboration and intense dedication to Varroa-mite research between qualified Institutes of EC-Member countries. At present the main research Institutes involved are those of Belgium, Denmark, Federal Republic of Germany, France, Great Britain, Greece, Ireland, Italy and Netherlands (Tab. I), while Portugal and Spain, recent full Members of the European Communities, are both on the point of becoming associated with the joint work.

The European activities on Varroatosis are now in progress in the framework of a five-year programme (1984-1988) implementing Council decisions on plant protection research. These actions develop through common activities and coordinated activities.

Selected Institutes, chosen following a call for tender which found many worthy responses, are participating in the common activities, supported by the CEC with financial help of up to 50% of the cost envisaged for the research. Seven EC-Countries are more directly involved in the research on Varroatosis on a contractual basis. These studies are carried out in close collaboration between the countries involved and I am pleased to stress at this meeting the very important role of the German Institutes which are very throughly developing lines of enquiry together with Belgian, English, Greek and Italian Institutes. No less important however are the collaboration and agreements between other Community countries such as Great Britain with F.R. Germany and Netherlands, France with Italy, etc., as well as, logically, between the various Institutes of the country itself.

Of the many research activities under way, we may particularly mention those which deal with the way in which the mite acts on the bee, such as parasite-host endocrine interactions and sensory signals; its diffusion as a natural pathway into a bee colony; the reaction of the host with selection of Varroa-resistant bee genotypes; specific techniques such as the development of artificial rearing for V. jacobsoni, quantitative diagnoses of Varroatosis, standardization of normal tests and methods of enquiry in the laboratory and in the field; research into efficient therapeutic means also making use of polyvalent methods such as Varroa attractive or repellent baits, the use of pathogens, physical treatments, chemical products, etc.

Together with joint actions, there are coordinated activities which develop above all through the organization of meetings of experts, seminars and workshops, and by the exchange of researchers between Institutes of various countries which are working in the same sector. These meetings allow an open discussion and setting out the various problems being studied, an exchange of information by the direct presentation of the research underway and the results obtained, a checking of the aims to be achieved with the definition of the best way of progressing to reach the desired goal quickly. In fact, in a nutshell, they allow a useful moment of reflection and of immediate reciprocal transfer of knowledge. The exchange of researchers helps to contribute to a better direct agreement and collaboration between experts of different Countries who are working in the same field and at the same time to a rapid passing on of techniques and research methods, which are useful for an agreed and more precise knowledge between partners.

To these actions one should add the rapid publication and diffusion of the results of meetings and researches, through specific proceedings or reports.

There is no doubt that in the field of Varroatosis the CEC is very active, based on the understanding which exists between all the components of the bee-keeping world and the spirit of harmony which inspires both producers and researchers.

Our meeting should be seen in this context and certainly its implications will have considerable effect on the future development of European activities intended to solve the serious and urgent problem of Varroatosis.

Welcome address on behalf of the German beekeeper's association (DIB)

E.Schieferstein
President of the German Beekeeper's Association

In the name of the German beekeepers I welcome you to this important meeting.

I hope that you have brought many new information and ideas which will result in a fast progress of your work.

As you may know, the beekeeper in the Federal Republic has a long experience with Varroatosis. The German Beekeeping Association supported the European research programme from the very first beginning.

Dr. F. Gnadinger who was the Association's President at the time played an important role in its initiation. This was a consequence of a long term policy of the Association to base the strategy of the control of Varroatosis on the three following points:

1. development and application of scientific methods;
2. participation of the gouvernment's veterinary service in all control campains;
3. cooperation of the Association and its organisations.

Even though experiments of private beekeepers were undertaken, the basis of official actions remained always professional scientific research.

The German beekeeper hopes and trusts in scientific knowledge and methods, which will guarantee the existance of beekeeping and honey production in the Federal Republic.

The European research programme offers a chance to work independently for this goal. The financial support allows to concentrate on the scientific problem. Commercial aspects of substances or methods must not gain priority. In this programme the European beekeeper is your 'employer'.

The German Beekeeping community has realised that control of Varroatosis without chemicals is not feasible yet. Nevertheless the following demands for the future are important:

1. minimal danger of contamination of bees wax and honey;
2. minimal disturbance of the biological cycle of the colony;
3. integration of control methods into beekeeping methods;
4. development of non chemical methods without considerable extra costs or work.

I know that it is easy to put foreward these demands. However, it will be very difficult and hard to reach those goals.

The German beekeepers know that they have to be patient with you - the scientists. Therefore, it is important to keep them informed on your work.

The problems and the progress of your work should be published in journals and in a form which beekeepers can read. This is my special request to you all.

I wish all success for your conference.

Session 1
Current situation of Varroatosis in European countries
Round table on Varroatosis in Europe

Chairmen: R.Cavalloro
D.A.Griffiths

Summary of the present status of Varroatosis in Europe

R.Cavalloro
Commission of the European Communities, Joint Research Centre, Ispra, Italy
D.A.Griffiths
Ministry of Agriculture, Fisheries and Food, Slough Laboratory, Slough, UK

A review of the present situation of the Varroa jacobsoni Oud. infestation in European Countries (Fig. 1) put in evidence the follow:

- Belgium (Prof. O. VAN LAERE) = V. jacobsoni is found in the regions bordering Germany and Netherlands. Apparently the spreeding of the mite is relatively slow. There were no damages due to Varroatosis reported up to data.

- Denmark (Dr. H. HANSEN) = There were no reports from Denmark yet. But the mite has reached the border in Schleswig Holstein recently. Precautions will be taken to prevent Varroa from entering the Danish islands.

- Federal Republic of Germany (Dr. F. GNADINGER) = The treatment of Varroatosis is practised with official support of the veterinary services. Mass losses of bee colonies occured only in the first year (1981-82). But still there are local damages. The general pollination of agricultural crops was not endangered, but there were some indications of failure of pollination in an apple growing region. Altogether no reliable estimation of the economic damages caused by V. jacobsoni in Germany is available yet.

- France (Dr. R. BORNECK) = There were no major losses of colonies in the eastern part of the country. But the situation in some southern areas and at the Mediterranean is more dangerous because the colonies have sealed broods throughout the year and no control method is available which allows to destroy the mites in the brood cells. French beekeepers were controlling Varroatosis by applications of Amitraze mainly, but remarkable damages were reported in spite of multiple treatments with Amitraze. A resistance of V. jacobsoni against Amitraze might be a cause for the failure of this treatment.

- Great Britain (Dr. D.A. GRIFFITHS) = Varroa was not found in the country yet. All measurements to prevent importation of the mite were put in operation. In case of occurence Folbex VA will be used for diagnosis.

- Greece (Prof. L.A. SANTAS) = Substantial losses of colonies due to Varroatosis which were common in the first years after the invasion of the mite do not occur any more. In Greece Malathion is generally used

Fig. 1 - <u>Varroa jacobsoni</u> Oud.: distribution and spread-Europe May 1986.
<u>Arrows</u>: indicate spread of Varroatosis through migratory beekee-
ping, <u>Asterisks</u>: indicate direct movement between two countries
by long distance transportation, <u>Cross hatch</u>: countries still
free from infestation.

to control <u>V. jacobsoni</u>. Beekeepers apply it after honey harvest and
reduce the number of mites to minimal level. Though Malathion is in
general use now for more than three years no reduction in its efficacy
against <u>Varroa</u> was observed.

- <u>Ireland</u> (Dr. H.D. HUME) = A survey of the extention service on a limited number of colonies was carried out. No <u>Varroa</u> was detected.

- <u>Italy</u> (Prof. F. FRILLI) = After its first report in the country far in
the North near Trieste, <u>V. jacobsoni</u> spread with an enormous speed due
to migratory beekeeping. Today the mite is found in many areas including Sicily and Sardinia. Folbex VA is generally used.

- <u>Netherlands</u> (Dr. A. DE RUIJTER) = In spite of quarantine measures
<u>Varroa</u> has spread throughout the country. The main treatment applied

12

was Folbex VA and recently Perizin. Due to the activities of the government extension in personnel and to the beekeepers, losses of colonies did not happen yet.

- Spain (Dr. A. GOMEZ PAJUELO) = First infestations were found recently near the French border. But much more serious were reports of V. jacobsoni near Alicante, in the surrounding of a settlement of German tourists. This is one of the major beekeeping areas of Spain and many thousands of colonies have migrated into the area. So, the mite might spread in Spain very fast, like in the other European countries with migratory beekeeping.

- Yugoslavia (Prof. J.M. KULINCEVIC) = V. jacobsoni entered the country already in 1974. It spread rapidly and besides some regions in the North the mite is present everywhere in the country. Beekeepers used Danicropa, Phenothiazin and some preparations of Amitraze for treatment of colonies. The development of the mite and the losses of colonies followed a specific pattern. In 1984 many colonies were destroyed or badly weakened by Varroa. Last year, 1985, the colonies developed well and the losses were replaced. For 1986 the observation of mass mite reproduction indicates that again damages are to be expected.

The round table discussion based on topics which the earlier statements showed up to be common to a number of countries.

Factors influencing the spread of V. jacobsoni and the rate of increase of mite population were discussed. Apparently beekeeping practice and climate represent significant factors. Information was brought out regarding current losses. The mites reproduce in strong colonies with plenty of brood until they cause heavy damages which reduce the colony strenght to a minimum. At this point the mites find not enough bee brood for their own reproduction. If few colonies survive then the mites are reduced and colonies build up rapidly. The Varroa reproduction catches up and the cycle starts again.

The discussion also brought up for the first time deleterious effects upon the pollination of agricultural crops.

Failures in control caused by resistance of Varroa to a certain acaricide was introduced in the discussion. These problems might play an important role in the future and can easily endanger all progress made in chemotherapy so far.

13

The spread, tracing and control of *Varroa jacobsoni* in Belgium

L.De Wael, M.De Greef & O.Van Laere
State Research Station for Nematology and Entomology (CLO) Ghent, Belgium

During recent years, the Varroa parasite has spread considerably in our neighbouring countries; in the Netherlands, the Varroa infection has moved to within a very short distance from our northern frontier and along the German and French frontiers, too, there is a threat of Varroa infection. The situation is depicted schematically in Fig. 1.

In Belgium, the first foci of Varroa were discovered in 1984, in the neighbourhood of the german and dutch border. Several other foci have been identified since then, as depicted in Fig. 2.. Officially, a total of twenty-three foci of Varroa are now known.

Measures for control of Varroa in the foci ascertained

Directives for control of bee diseases have been laid down by the Veterinary Inspection Service of the Ministry of Agriculture :
- Varroatosis is classified under infectious diseases, which implies on obligation on everyone to report it.
- The treatment of varroatosis in an area is carried out under the responsibility of the Veterinary Inspector of the Ministry of Agriculture.
- The treatment is carried out by an Apiculture Assistant, who makes use of chemical methods. Those Assistants are bee-keepers specially trained for dealing with diseases.
- The laboratory and treatment costs are borne by the State.

Measures for preventing further spreading

- At a radius of 5 km around the foci of infection, a protective area is established. The movement of bees across the protective cordons is prohibited.
- Any bees that must be moved, must be accompanied by a health certificate. Fur that purpose, a debris-test is required.
- To each bee-hive located on ground not adjoining the habitation of the bee-keeper must be attached, clearly visible, the name and address of the bee-keeper, together with the health certificate.
- For the sale of colonies of bees, queen-bees or swarms, a health certificate, issued by the Veterinary Inspector, is required.

Those measures clearly are of a temporary nature. Protective cordons and prohibition of travel are useful only as long as the varroatosis remains limited to a few foci, such as has so far been the case.
- Last year and again this year, the Ministry of Agriculture organised a general search operation for varroatosis ("Screening").

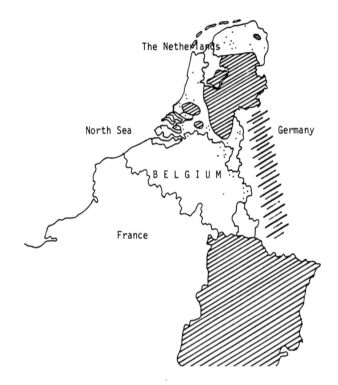

Fig. 1. - Situation 1986.

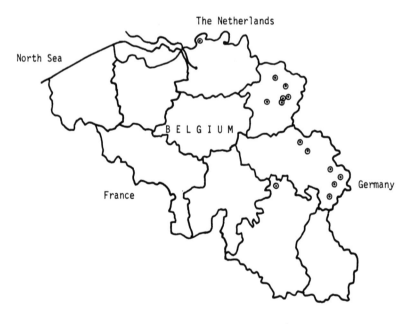

Fig. 2. - Varroa foci in Belgium up to september 1986

16

Screening organisation

In view of the situation of the infection in the Netherlands, Germany, France and in the eastern part of Belgium, it appears to be necessary to carry out a thorough screening programme, for the purpose of detecting the initial phase of the disease as early as possible. In that way, bee-keepers can be allerted in good time and a massive loss of colonies can be prevented.

For practical reasons, Belgium is subdivided into twenty-five districts of the Stock-breeding and Veterinary Department. Each district is under the responsibility of a Veterinary Inspector.

As regards apiculture, that Inspector is legally assisted, by one or several Assistants. In addition, through the Assistants, agents in apiculture are also involved.

It was proposed that, in the latter part of 1985, some 100 bee-hives per veterinary district should be sampled.

This screening programme was based on the voluntary collaboration of the bee-keepers, who contact the Veterinary Inspector. The samples are taken by certified Apicultural Assistants and agents.

The samples tested consisted, for the greater part, of bottom debris, collected from the middle of August to the end of September. In a number of cases, fumigation with Folbex VA was carried out first.

The debris was dried for 24 h at 55°C. After covering with technical ethanol, dry mites float on the surface. If no mites appear, distilled water is added. The density of the solution is thereby increased and any mites still remaining or damaged mites then float to the surface.

Table 1 summarises the number of samples tested in the Dutch-language provinces, per district

Table I. Number of samples tested

Province	District	Number of samples
West Flanders	I	135
	II	64
	III	100
East Flanders	IV	135
	V	15
	VI	20
Antwerpen	VII, VIII	190
	IX	52
Limburg	X	24
	XI	34
Brabant	XII	) 33
	XIII	)

In table 2, the number of bee-keepers who participated in the screening tests, is placed opposite the total number of bee-keepers per Province.

Table II. Participation percentages

Province	Number of members	Number of bee-keepers participating	Percentages
West Flanders	873	96	11.0
East Flanders	991	48	4.8
Antwerp	1716	135	7.9
Limburg	1080	52	4.8
Brabant	813	15	1.4
Total	5473	346	6.3

These screening tests are being continued in 1986. The total number of offers of participation is 1483 to date. Consequently, 213 bee-keepers are collaborating in the detecting campaign.

Conclusion

In 1985, none of the districts achieved the stipulated collaboration. It was the intention, with the screening tests, to sample 100 bee-hives per district. This represents 24 % of the bee-hives. Last year, only 6.3 % of the bee-keepers participated. This year, too, the number of inquiries is below that of the collaboration aimed at.

Perhaps, the bee-keepers are still insufficiently aware of the Varroa problematic, as long as Varroa is not present in the immediate neighbourhood or on their own ground.

The beekeeper and Varroa-disease

F.Gnädinger
Deutscher Imkerbund, Stockach, FR Germany

There is no other subject about which has been written
so much in bee-journals as about Varroa-disease. It is an in-
exhaustible theme with facts and opinions up to wish-dreams.

1. Characterization of the bee-keeper

What are the characteristics of bee-keepers? The answer
is: they are coming from all kinds of professions, from all
classes of people, from all education levels, from the simple
man up to the high ranking university professor.

You must also know that one bee-keeper attends to 13 -
15 colonies on an average and that he is bee-keeping in about
99 % as an additional occupation. Apiaries can have between
3 and 300 colonies. For nearly all bee-keepers economic reasons are conspicuous for bee-keeping; to me only very few are
known occupying themselves with bees as a pure hobby.
There are only few professional apiarists living only
by bees; income is too uncertain.
The bee-keeper will absolutely be successfull: reality
and wishes in judging the colonies often differ ("bee-keeper's latin").
The honey normally is being sold directly to the consu-
mer.These are also a special circle of persons. They absolutely want to have a natural product and are willing to pay
a higher price for it. It is an event of confidence.

2. Training

The first reports on Varroa-disease were contradictio-
nary. They did not allow either the scientists or the practical bee-keepers a safe judgement.
First of all, it was the "Insitut für Bienenkunde" in
Oberursel who had to engage in research about Varroa. Later,
nearly all bee-scientists were rushing on this fascinating
field.
Bee-keeper's Associations, especially the "Deutscher
Imkerbund" induced by means of circulars, courses of instructions, performances and periodicals a training as it didn't
happen in the history of the bee-keeping organisations. From
this a close cooperation between scientists and associations
resulted. It was incredible for many people to hear that colonies suffering from Varroa-disease and not being treated
would die, under European conditions. They thought it would
not be as bad as told. Only after having seen the catastro-

phy at their own colonies they began to believe in the danger of this disease.

The sometimes immature remarks, even of experts, made many of the bee-keepers insecure. This even happened at the Congres of German speaking bee-keepers in St. Gallen/Switzerland in August 1986. After many discussions, bee-keepers asked themselves: "What's correct now?" They expect from the scientists a conception for a practical, successfull control. If this will not be possible the bee-keepers will have to look for their own way.

3. Detailed problems

3.1   The application of pharmaceutical products in bee-colonies mostly is very unpleasant for the bee-keeper.
You can divide into:

      beekeepers who believe to be able to control the Varroa-disease only by using biotechnical methods. In the meantime we have a number of variations of the biotechnical basic methods, which are recommended. The application of medicaments is refused.

      beekeepers who instruct themselves more in scientific direction, who accept a little quantity of unobjectionable remainders of medicaments in honey or wax - though mostly unwillingly.

      beekeepers preferring a combination between the biotechnical methods and the medicamental methods. They believe this to be the best way for the near future.

The apiaries differ by size of colonies, honey-flow, management (e.g. beekeeping at a fix place, migratory beekeeping with a greater number of colonies, equipment for transport) considerably.

Essentially, beekeepers have multiple-storey hives, but in different types. In addition to this there are still side-opening hives e.g. in the Black Forest, with their own managements.

3.2 Concerning medicaments
    Beekeepers have the following desires:
    1. good efficacy, no residues
    2. cheap
    3. simple application
The present, licenced medicines do not fully correspond to these 3 demands.

FOLBEX-VA-Neu (Brompropylat)has a good efficacy, but does not penetrate through the comb-cappings to destroy the mites during the multiplication. The application is complicated. Residues in wax are possible, the medicine is expensive.

PERIZIN (Cumaphos) has a good efficacy. It does not penetrate through the cappings with the same result as described at FOLBEX-VA-Neu. The application is very simple. Residues in wax are possible, the medicament is expensive.

The ILLERTISSER MITE PLATE (60% acid formic) has a very different efficacy. Sometimes, bees leave the hive after treatments and form a swarm at the entrance. Sometimes we have losses of queens. The medicament penetrates a little through the cappings and destroys the progeny of the mites. It is not soluble in wax but in honey. The application is simple but restricted on the time after extraction during August and September. The plates are cheap.

PERIZIN is the medicament which presently fulfills the wishes of the beekeepers most.

There also are bee-hives where only FOLBEX-VA-Neu can be used because it effects by smoke.

Now a word concerning the not licenced medicament AMITRAZ: it is uncomprehensible for the beekeeper that a medicament with a good efficacy and a very cheap price, being allowed for use in France and other countries, may not be applied in Germany. The same medicament should have the same judgement ind the same situation in all countries of Europe.

Futhermore: The beekeeper fears the cumulation of residues of medicaments in the wax. Therefore, he desires the destruction of the chemical compounds in the wax by melting with transformation for comb-foundations. In Germany, therefor the wax must be heated at a temperature of 180°C for a time of 30 minutes. For the beekeeper it is difficult to understand that this temperature shall not be sufficient to destroy the medicaments.

The president of the German Beekeeper's Association, Dr. Schieferstein, recently said: "It is impossible that there still are residues in the honey!" This, of course, was a political statement which does not correspond to the reality but represents a general desire.

3.3   It was not successfull to control the Varroa-disease by means of state-legislation. Varroa-disease was not eradicated Only the spreading was slowed down. Therefore, it is understandable that Veterinary Services want to take out the control of Varroa-disease from the state-regulations. The Beekeepers' Associations as well as most of the beekeepers want to liberate the measurements but at same time no loss of the assistance of the state-services e.g. financing experts, scientists in institutes, diagnosis etc. and last but not least help at buying expensive medicaments in pharmacies, which are subject to precriptions.

State-regulations should be adaptable to the different situations. There are no dificulties when Veterinary officers are beekeepers themselves with own experiences. Occasional egoistic claims of beekeepers having no intention to pay regard to their colleagues with not-infested colonies, cannot be fulfilled.

4.   At the last meeting in Saloniki, Professor Cacalloro proposed the registration of the losses and damages by Varroa-disease. I was worried about that because we sometimes had regional information on insufficient pollination, especially of apple-trees. The farmers believe that only in very few cases the reason could be a lack of pollination. They

21

mostly think it to be a backstroke of chill. But we don't
have sufficient information to registrate the losses and damages in pollination.

By the intensive training, the beekeepers have learned
to fight the Varroa-mites, so that hard losses at bee-colonies only happen at the first intestation. Therefore it is
difficult to fulfill the desire of Professor Cavalloro.

Certainly, each year we have losses of bee-colonies by
Varroa-intestation. I estimate these losses by last winter
and this year to 20-25 %.

Current status of Varroa-disease in Greece

L.A.Santas
University of Agricultural Sciences, Athens, Greece

Summary

Now Varroa disease in Greece is present in the whole of the mailand and in most of the islands. Kea, Syros, Seriphos, Kythira, Antikythira are the only islands unifested.

During the first three to four years following the discovery of that serious disease in Greece (1978), and mainly during the period 1981-82 the bee-colonies had been heavily damaged.

Later on and after the use of some acaricides the problem has resticted and now is not so severe.

The main chemical for the control of Varroa disease in our country is Malathion. In Greece this pesticide is applied as dust below 1‰ a.i. between frames or over the combs.

1. Spread

Varroa disease a dangerous pest of bees, is known to be causcd by the ectoparasitic mite Varroa jacobsoni Ouds. This pest, is rather a new invader in our country. It was first recorder near Greek-Bulgarian borders in April 1978(5). Sortly (late in the same year), after the discovery of the mite in Greece, the beehives in many other places in Thrace, Macedonia and Central Greece were found to be infested(3).

Obviously the mite was introduced to Greece from Bulgaria and Yugoslavia where the disease was already known from 1967 (Bulgaria) and 1976 (Yugoslavia) (1,8).

Isolated cases of beehives affested by Varroa mite were also observed in Southern Greece during the winter of the next year (6).

After that, Varroa disease was spread very quickly and soon it was present in the whole of the mainland and most of the islands (Fig.1).

The islands of the Eastern Aegean were infested earlier than other islands which indicates that the infestation was introduced there from Turkey(2).

Now Varroa disease in Greece is present in the whole of mainland and most of the islands. Kea,Syros,Seriphos,Kythira,Antikythira are the only islands unifested (Fig.1).

There are no accurate data in our country concerning the economic losses from Varroa disease in the bee colonies. However during the first three to four years folowing the discovery of that serious disease, April 1978 and mainly during the period 1981-82 the bee-colonies had been heavily damaged and many apiaries in various places were destroyed by this mite.

The Greek Beekeepers Association estimated that 20-25% of the country's bee colonies have been heavily damaged and some times totaly destroyed by this disease(7).

23

Later on and after the use of some acaricides the problem has restricted and now it is not so severe.

Malathion as powder is considered and used in Greece as bee medicine. It is the main chemical for the control of Varroa disease in our country for many years and it is used with great success as powder, in concentration below 1‰, applied mainly by hand between frames or over the combs (2,4,7).

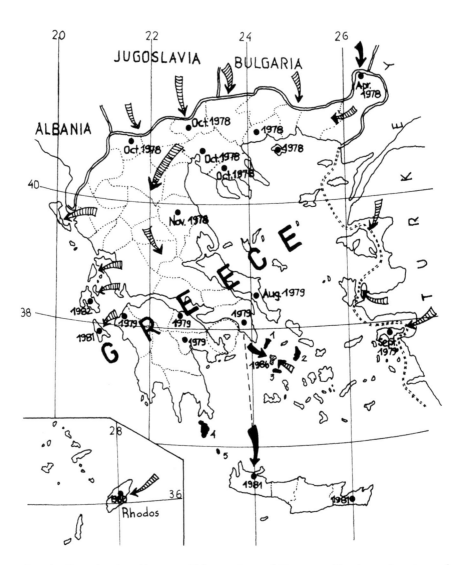

Fig. 1. Map showing the possible routes of Varroa mite invasion to and within Greece. Islands in Black colour are unifested. (1 = Kea, 2 = Syros, 3 = Seriphos, 4 = Kythira, 5 = Antikythira).

2. Manner of spread

The spread of the mite within a country depends on various factors. In Greece the spread has been very quick. Three years after the first disco ry of the mite in the Nothern part of the country (1978) it has been sp almost all over the mainland even in the most of the islands (Fig.1).

Although, during the first years of the mite discovery the govemen estallished severe regulations to prevent the spread of the mite, these re not worked well. The beekeepers would not wanted or ignored or they ld not believed in the severity of this pest. Thus these favoured a rap spread of the species.

An important factor was the strong migratory activity of the beekeep in Greece. The beecolonies are transfered every year to various bee pas

Thus in citrus groves and thyme areas in spring, in fir forests in M June, in cotton and glover plantations in summer and in pine forests in gust, September and October. This expanding practice of migratory beeke ping encouraged the contamination of unifested areas.

On the other hand, but in less extend the stray swarms or single bee has distributed the mite in new areas.

3. Conclusions

The mite Varroa jacobsoni is continuing to spread in Greece. All bee lonies of the country out of ones in the islands of Kea,Syros,Seriphos thira,Antikythira are infested.This year, September 1986,a few beecolor of Kythnos island were found to be infested by the mite (Fig.1).

The use of Malathion as powder in low concentration, below 1‰, keel the mite population in a low economique lever.

REFERENCES

1. DE JONG,D.,MORSE,R.,EICKWORT,G.,1982. Mite pest of honey bees.Ann.R Entomol. (27):229-252.
2. EMMANOUEL,N.G., SANTAS,L.A., TABOURATZIS,D.C.,1982. Varroa disease its control in Greece. VI Intern.Congr.of Acarology,Edinburg 5-11 S 1982, pp. 1099-1105.
3. IFANTIDIS,M.D., THRASYVOULOU,A.,1978.Varroatosis.A new mite, attack the honeybee in Greece and its effects on beekeeping.GEOTECNICA 4:3
4. IFANTIDIS,M.D., 1980.Malathion als Kontaktmittel zur Bekäpfung der roamilbe pp. 144-149 from Diagnosis und Therapie der Varroatose. Ap Publis. House, Bucharest, Romania.
5. PELEKASSIS, C.D., SANTAS,L.A., EMMANOUEL,N.G., 1978 . Varroiki aka sis. A new for Greece serious honeybee disease. "Melissa" 1-7 pp., ns (in Greek).
6. PELEKASSIS, C.D., SANTAS,L.A., EMMANOUEL,N.G., 1979 . Varroa disea Greece (Distribution - Morphology - Control measures) XXVIIth Inte: Congr.of Apicult.Apimondia, Athens, Greece, Sept. 14-20, 1979, pp. 365. Apimon.Publ.House, Bucharest, Romania.
7. SANTAS,L.A., 1983. Varroa disease in Greece and its control with M thion. In proc. of Meeting of experts group "Varroa jacobsoni Ouds fecting honey bees:Present status and needs" Wageningen, 7-9 Februi 1983, pp. 73-76.
8. SPITZER,M.,SULIMANOVIC,D., 1986.Control of Varroa disease in Yugos "Varroa Workshop" Feldafing/W.Germany 24-26 Aug.1986.(Abstacts p.2

Varroatosis – The Irish situation

H.D.Hume
Department of Agriculture, Dublin, Ireland

During the Spring of 1983 the Department of Agriculture requested bee-keepers to submit one sample of hive debris per apiary for screening for Varroa jacobsoni. Ninety three samples were received and examined by the Bee Disease Diagnostic Service of the extension service ACOT. Varroa jacobsoni was not found in any of the samples.

Although a specific survey for Varroa has not been conducted since 1983, a significant number of beekeepers continue to avail of the Bee Disease Diagnostic Service. During 1984 and 1985 a further 82 and 36 samples of hive debris were examined and found to be free from the causal organism of Varroatosis.

In addition the Service receives annually a number of samples of bees and brood comb for examination for diseases known to be present in the country. Tables 1 and 2 give the relevant data for the years 1983, 1984 and 1985.

Table 1. Number of bees examined by the Bee Disease Diagnostic Service.

Year	No. of Bees	No. of Colonies
1983	40,352	1,852
1984	37,936	1,932
1985	46,670	2,472

Table 2. Number of combs with varying degrees of brood examined by the Bee Disease Diagnostic Service.

Year	No. of Combs	No. of Colonies
1983	223	223
1984	226	226
1985	300	300

These samples were randomly checked for the presence of V. jacobsoni and were found to be negative.

As it is assumed that the disease is not present in the country the measures enacted during 1980 i.e. Bees (Regulation of Import) Order, 1980 (S.I. No. 161 of 1980) continue to remain in force. During the period June 1981 to May 1986 licences to import bees were not issued. Since June 1986 two licences were issued authorising the importation of queen bees only and from Northern Ireland only.

Current situation of Varroatosis in Italy

F.Frilli
Istituto di Difesa delle Piante, Università degli Studi, Udine, Italy

Varroa jacobsoni was found in Italy for the first time in 1981 in Friuli-Venezia Giulia (2,4). Much to the amazement of many the following year a substantial centre of infection was discovered in central Italy on the boundary between Tuscany and Latium. The levels of infestation were much higher than they had been in Friuli-Venezia Giulia (1). Surveys carried out during the following years showed how the parasite could be found now in one Region and next in another without any gradual spread.

Since the investigations on the finding of Varroa have been very intense from 1982 onwards, it is easy to show how in the various regions of Italy infestations are caused by the transport of infected apiarian material or the buying and selling of the same and not by the spread of the mite. It is possible now to ascertain how in four years fast all the twenty Italian regions were infested (fig. 1).

The first infestation noted in the north east of Italy (Friuli-Venezia Giulia) convinced the national health authorities to impose extremely strict regulations on the movement of beehives and on the destruction of apiaries. If on the one hand the transport of hives into the nearby Veneto Region was prohibited and so the most dangerous method of Varroa diffusion blocked, on the other the uselessness of destroying entire apiaries, when the infestation level is high, as a means of prevention was demonstrated.

As far as the speed with which this parasitic disease spreads is concerned, it is interesting to note a study carried out in the Institute of Plant Protection at Udine University to check how the Varroa moved from the initial centre of infestation to the surrounding zones.

Having come from Yugoslavia and been detected in June 1981, the mite was actively sought in the following months.

From a survey carried out in the spring of 1982 it was concluded that only the province of Gorizia appeared to have an infestation of light to medium intensity, but one year later the provinces of Trieste and Gorizia were found to be seriously infested, while a border area of Udine was weakly infested. From samples taken at a distance of one year (spring 1984) and two years (spring 1985) it could be seen that by that time the infestation had spread over the entire region (3). Looking at a map it can be concluded that the speed of diffusion in our region was about 20 Km per year, in other words near enough to that mentioned by Grobov (5).

The progressive worsening of the infestation, above all in 1983, led to the loss of numerous apiaries not adequately protected by chemical

Fig. 1 - Years in which **Varroa jacobsoni** was first detected in each region of Italy.

methods. Now "cohabitation" with varroatosis has been achieved, and it is no longer unknown as bee-keepers are interested in the problem. A type of natural selection has taken place, and is still in progress in some areas, where insufficiently motivated bee-keepers succumb.

A final warning: the increase in the "american foulbrood" in apiaries of various areas, with easily imaginable consequences.

References

1. ACCORTI, M., DE PACE, F. (1983). **Varroa jacobsoni** Oud. nell'Italia centrale. L'apicoltore moderno, **74**, 3-6.

2. BARBATTINI, R. (1981). Presenza di **Varroa jacobsoni** Oud. in territorio italiano. L'Informatore agrario, **37** (31), 16769-16770.

3. BARBATTINI, R. (1986). La diffusione della varroasi in Italia. La città delle api, **21**, 27-31.

4. FRILLI, F. (1983). **Varroa jacobsoni**: The situation in Italy. Proc. Meeting EC. Expert's group, Wageningen 7-9 Febr. 1983, pp.15-18.

5. GROBOV, O.F. (1977). La varroase des abeilles, in: La varroase de l'abeille mellifère, Apimondia, Bucharest, 52-98.

Present situation of *Varroa jacobsoni* Oud. in Italy

R.Prota

Istituto di Entomologia Agraria, Università degli Studi, Sassari, Italy

Summary

The first official notification of **Varroa jacobsoni** in Italy came in 1981 from areas near Yugoslavia. From then on, reports of the mite presence have multiplied, eventually regarding all regions. Sardinia and Sicily appear seriously affected.

The information reported here forms part of the observations made by the various research institutes collaborating in our country on an Api-cultural Research Project supported by the Ministero della Pubblica Istru-zione.

Varroa jacobsoni was first officially notified in Italy from areas near Yugoslavia in 1981 (2). From then on, multiplying reports of the mite regarded many other regions: Friuli-Venezia Giulia, 1981 (2); Lazio, 1982 (1); Tuscany, 1982 (1); Umbria, 1982; Sardinia, 1983 (4); Lombardy, 1983 (5); Campania, 1983; Sicily, 1983; Veneto, 1984; Emilia, 1984; Puglia, 1984; Liguria, 1985; Marche, 1985; Abruzzo, 1985; Piemonte, 1985; Basili-cata, 1985; Calabria, 1985. High infestations were reported from Friuli-Venezia Giulia in the north, Tuscany and Campania in the central region, and Puglia in the south. Southern areas as well as Sicily and Sardinia show the heavier infestations, with environmental conditions favouring the development and diffusion of the mite (Fig. 1). The present situation in the latter region is particularly serious; many of the apiaries are so badly infested that colony losses are as much as 65% (Tab. 1). After the issue of Ministerial Circular Notices at national level, many Regional Governments have sought to restrain diffusion of the pest by way of legal provisions aimed at localizing the infested areas and prohibiting transfer of the colonies unless accompanied by a clean certificate of health (3).

Research Activity

Organizations such as the CEC, National Council for Sc. Research, the MPI and several Regional bodies are now financing research by many scientific institutes not only into diffusion of the mite but also into diagnostic and therapeutic methods. Remedial measures used in practice include the frequent use of bromopropylate, in many cases nebulization with amitraz, and intervention with substances of vegetable origin. In the experimental field numerous trials are in course using diverse methods and chemical preparations. Particularly interesting are those introducing

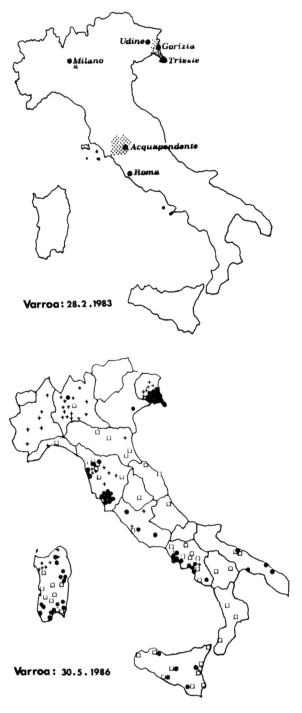

Fig. 1. Infestation diffusion in Italy from 1983 to 1986.

new systemic agents (e.g., coumaphos, cimiazole) into the colonies periodically, or fumigation with bromopropylate from flight entrance. Control with natural substances, such as menthol, thymol, and eucalyptol is being tried, and also synthetic preparations (legally allowed in other countries) not excluding the pyrethroids; but the results have not always been satisfactory and are often contradictory.

It is, perhaps, hardly the case to mention that the greatest amount of experimentation occurs in the northern apiaries where, however, infestation has not reached the critical levels beginning to be manifest in our islands. This extremely dangerous state can be attributed to the high rate at which the mite multiplies and the acceleration of its cycle at the expense of the female brood, but also to the particular habit of our native **A. mellifera** to spread its oviposition over a relatively long period (fig. 2).

Considering the present situation, we feel that the interest and incentives of the CEC, reserved up to the moment largely to northern mainland regions (some of which are free from **Varroa**), could be more strenuously directed towards the apiculture of the South, which is also a highly important factor in fruit and vegetable growing and conservation of the natural flora.

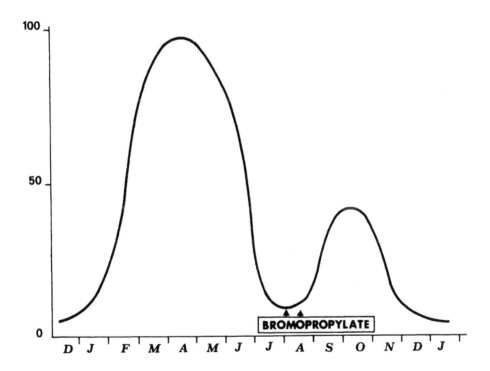

Fig. 2. Trend of brood egg presence during the year in northern Sardinia areas without summer flowering plants; (↑___↑) recommended period for chemotherapeutic intervention.

Tab. 1. Italian regions where **Varroa jacobsoni** Oud. has been noted and estimated percentaged of colonies lost.

REGIONS	YEAR	%
FRIULI-VENEZIA GIULIA	1981	50
LATIUM	1982	55
TUSCANY	1982	40
UMBRIA	1982	50
SARDINIA	1983	65
LOMBARDY	1983	-
CAMPANIA	1983	50
SICILY	1983	60
VENETO	1984	-
EMILIA	1984	-
APULIA	1984	40
LIGURIA	1985	-
MARCHES	1985	-
ABRUZZI	1985	-
PIEDMONT	1985	-
BASILICATA	1985	45
CALABRIA	1985	-
MOLISE	?	-
TRENTINO	?	-
VAL D'AOSTA	?	-

REFERENCES

1. ACCORTI, M., DE PACE, F. (1983). **Varroa jacobsoni** Oud. nell'Italia centrale. L'apicoltore moderno, 74, 1, 3-6.

2. BARBATTINI, R. (1981). Presenza di **Varroa jacobsoni** Oud. in territorio italiano. L'informatore agrario, 37(31), 16769-16770.

3. FRILLI, F. (1983). **Varroa jacobsoni**: The situation in Italy. Proc. Meeting EC Experts' Group, Wageningen 7-9 Febr. 1983, pp. 15-18.

4. PROTA, R. (1983). Sulla presenza di **Varroa jacobsoni** Oud. in Sardegna. Studi Sass., sez. III Ann. Fac. Agr. Univ. Sassari, Vol. XXX, 255-264.

5. SOMMARUGA, A. (1983). Circolare n. 143, 11.03.83, Laboratorio Apistico Regionale della Lombardia.

The situation of *Varroa jacobsoni* in Spain

A.Gómez Pajuelo, J.L.Molins Marín & F.Pérez García
Centro Experimental Agrícola y Ganadero, Diputacion de Cadiz, Jerez de La Frontera, Spain

Summary

Varroa jacobsoni Oud. appears in Spain in Albacete (September 1986), Alicante (July 1986), Ciudad Real and Cuenca (September 1986), and Gerona (December 1985).

In the most important area, Alicante, it was present at least two years before its official detection.

The infestation is more widespread than official data indicate. Detection and treatment is carried out with Folbex VA and Taktic. A quick field-method for diagnosis has been designed, which detaches the mites with 25% ethanol.

The official norms, which are very severe, force apiarists to hide any evidence of mites.

1. Areas of infestation

The presence of Varroa jacobsoni Oud. in Spain was first detected in December 1985, in Gerona, near the frontier with France.

In July 1986 an important centre of infestation was detected in Callosa d'en Sarrià (Alicante), near a housing estate largely occupied by Germans who keep bee-hives in their gardens, according to verbal reports by local beekeepers.

Hives have been detected with levels of infestation from 60 to 30% (1), which suggests that they have been infested for 2 - 3 years.

In the month from October to November some 12000 hives are gathered in this area, 80% of them seasonally migratory, for the blossoming of the medlar tree (2).

In August and September 1986 new areas have been found, officially, in Albacete, Ciudad Real, and Cuenca, with hives which had been some years before in Callosa.

Fig. 1 - Centres of Varroa infestation in Spain until September 1986.

All this would suggest that the infestation is much more widespread than the official data indicate (Fig. 1).

2. Detection and treatment, products and techniques

Most detection and treatment so far has been carried out with Folbex VA, the only product legally registered in Spain to be used on bees and which is distributed free by the Ministry of Agriculture, Fishing and Food.

In some areas Taktic is used, applied by micro-spray.

Both with Folbex VA and Taktic, greased paper is used on the base of the hives to detect the presence of mites.

Most Spanish hives are seasonally migratory, belonging to professional and semi-professional apiarists, who almost exclusively use Layens, of 10 - 12 frames with a fixed base and a small entrance hole of 8 x 1,5 cm.

Therefore, for the placing of greased paper, frames have to be removed, which is a slow and costly process.

To avoid this, a quick method of field-sampling has been designed, based on De Jong 1982 (3).

3. Legal norms

Varroatosis is a parasitosis which must compulsory be declared in Spain to the sanitary Authorities of each Autonomous Community (4).

In order to move bee-hives it is necessary to obtain a certificate of negative mite diagnosis, issued by an official laboratory, which is valid for two months.

Unfortunately, in some areas as extensive as Andalucia, Castilla, etc., there is only one official laboratory, and diagnoses of other laboratories are not considered valid.

Infested hives, including those with Varroa, were totally burnt at first. At present the swarm is destroyed (with SO_2) and the frames with young from highly infested hives are melted, the owner being indemnified with 3500 pesetas per hive (5). The rest of the apiary is immobilized and must obligatorily be treated, as must the apiaries of the same and bordering municipal districts.

Due to our climate, the blossoming times are very short, because of which migratory apiaries trapped in immobilized zones have escaped from the areas.

The most general idea of migratory apiarists, who are the ones who hold most of the bee-hives in Spain, is not to declare and to escape.

References

1. Regueira, D. - Consellerìa d'Agricultura, Alicante. - Personal communication (1986).

2. C.-Sociedad Cooperativa Apìcola de Espana, Valencia. - Personal communication (1986).

3. De Jong, D. - from Andrea Roma, D. & Goçalvez, L.S. - "A comparative analysis of shaking solutions for the detection of Varroa jacobsoni Oud. on adult honey-bees". - Apidologie 13 (3): 297-306 (1982).

4. Boletìn Oficial del Estado, p. 8-839 (8.3.1986).

5. Catalàn, M. - Servei de Producciò Animal, Valencia. - Personal communication (1986).

Session 2
Biology of *Varroa jacobsoni* Oud.

Chairmen: M.D.Ifantidis
J.Beetsma

Implied sensory signals in the honeybee-Varroa relationships

Y.Le Conte & G.Arnold
Laboratoire de Neurobiologie Comparée des Invertébrés INRA/CNRS, La Guyonnerie, Bures/Yvette, France

Summary

Factors involved in the phenomenom of distance attraction of varroa by honeybees are studied through 3 different approaches : an olfactory study to determine the implied molecules; studies on the role of heat; the effects of vibrations produced by honeybees on varroa's behaviour. To complete these studies, an anatomical approach is realized to determine the sensory organs involved in these relationships. More, the difference of attractiveness between emerging and older bees is studied.
It is found that a/ molecules are implied in these relationships, b/ heat and vibrations have an important effect on varroa's behaviour, c/ heat seems to be the most important factor among the other ones studied, d/ the varroa's temperature preferendum varies according to its age and the ambiant temperature, e/ the difference of heat produced by honeybees is a partial explanation for varroa's preference for older bees, f/ there is a sensory system on the distal part of the varroa's first pair of legs.

1.1 INTRODUCTION

In the perspective of an effective control of varroa disease and of a set up of integrated biological control, it is important to know perfectly varroa's biology and its relationships with honeybees.

Previous results showed that varroa is attracted by bees at the distance of 1 cm.(1). Besides, the bees release olfactory, thermic and vibratory signals. Therefore, we suppose that one or more of these signals are involved , at this distance, in the recognition of bees by varroa.

This work aims to determine the relative importance of each of these signals on varroa's behaviour :
 - Olfactory signals
     The role of semiochemicals is examined in order to determine volatile molecules produced by bees which could influence varroa's behaviour.
 - Thermic factors
     Workerbees produce heat and are able to maintain a temperature of 34°C in the brood (2). We have examined the relative attractiveness of heat on varroa, and the differential attractiveness of heat in comparison with olfactory and vibratory stimuli.
 - Vibratory signals
     Vibrations produced by a group of bees are tested.

In order to complete this work, an anatomical approach is realized to

41

Figure I - Four-tubes test olfactometer.

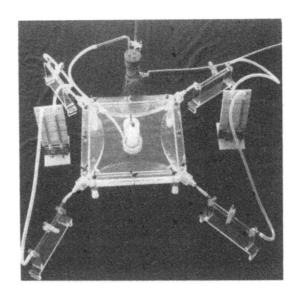

Figure II - Airflow olfactometer.

find what kind of sensory organs are implied in the detection of the different signals.

Moreover, we found a difference in varroa's attraction according to bee ages (1). The differential attractiveness of bees from different ages is observed as well as the effect of heat on varroa's distribution in the different categories of bees.

2. Materials and methods

2.1. Anatomical approach
A group of sensory organs on varroa's first pair of legs has been observed with a scanning microscope at the "Service Commun de Microscopie Electronique à Balayage de l'Universite Paris 6 et du C.N.R.S.".

2.2. Involved factors in the recognition of bees by varroa
2.2.1. Olfactometric study
        - Behavioural aspects
            Two kinds of olfactometers are used.
            * The first one, with passive diffusion of odours, is called the four-tubes test. It is made of a Petri dish with four holes drilled at the bottom. Glass tubes closed at their ends by a piece of tulle (figure I) are encased in each holes. Biological material is placed inside two of them, the two others are used as controls. Varroas are placed in the center of the Petri's dish. After a given time, the number of varroas on each piece of tulle is checked.
            * The second olfactometer is made up with active movement of odours by mean airflow. It has been built in our laboratory (from Vet (3)), to observe varroa's behaviour in different odorous airflows (figure II).

        - Chemical analysis
            Extracts of bees, brood, larvae, wax, etc.... are realized according to two different and complementary methods :
            * the trapping of the odours with glass noddle, porapak, active charcoal; then the extraction with solvents (hexane, dichloromethane, freon, etc...) and the concentration under nitrogen flow.
            * the extraction of cuticles with different solvents and the concentration under nitrogen flow.
    The different extracts are tested to determine their attractiveness on varroa.

2.2.2. Study of thermic factors

Varroa's thermopreferendum is studied with a special apparatus giving a thermic gradient of 1°C per cm. It is analysed in relation to the room-temperature and to varroa's age.
    The four-tubes test is used to make conspicious :
            - the relative attractiveness of heat by heating 2 tubes,
            - the choice of varroa between heat and alive workerbees and the choice between heat and vibrations produced by a group of 200 bees.

Varroa's position on the workerbees'bodies is noticed in function of room-temperature.

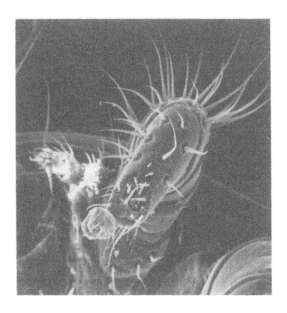

Figure III – Varroa jacobsoni Oudemans 1904 – Distal part of
the first tarsus. x 700.

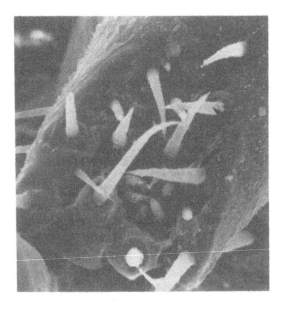

Figure IV – Varroa jacobsoni Oudemans 1904 – Sensorial plate of
the distal part of the first tarsus. x 2500

2.2.3. Study of vibratory signals

The four-tubes test is modified. Sounds produced by 200 bees arriving in the tubes are tested.

2.3 Differential attractiveness of workerbees in function of their age

2.3.1. Four-tubes test

The four-tubes test is used to experiment the attraction and then the choice of varroa between bees of different ages (emerging or more than 5 days old). The heat produced by the bees is measured with thermic sounding-lines.

The same experiment is repeated after the tops of the tubes are sealed by aluminium foil, so that the olfactory factors can not interfere.

2.3.2. Passage of varroas from young bees to older ones

* Passage of varroas in function of time :

40 parasited emerging workerbees are put in a special box of 6.5.4 cm.. 40 marked 5 days old unparasited bees are introduced into the box. After a determined time, the 80 bees are observed to calculate the percentage of varroas which have passed from emerging bees to olders. The experiment is repeated with different timings.

* Passage of varroa in function of bees ages :

The previous experiment is executed again and the period is fixed at 12 hours. Emerging bees are tested during each experiment in comparison with bees of known ages. Two kinds of tests are realized :
- passage of varroas from emerging bees to older ones,
- passage of varroas from older bees to emerging ones.

* Influence of room-temperature on varroa's distribution :

The same experiment is repeated with parasited emerging bees and unparasited 5 days old bees in relation to room-temperature and to the time the two categories of bees are together.

3. Results

3.1. Anatomical approach

This work make conspicious a structure at the distal part of varroa's first pair of legs (figure III). This structure is made up with an incurved sensory plate, with three kinds of sensilla :
- long sensilla (ts), placed at the periphery of the sensory plate (figure IV),
- a group of 6 smaller sensilla making a crown, with the top of the sensilla turned toward the center (cs),
- 2 short sensilla (ls) inside the crown of sensilla (figure V).

3.2. Involved factors in the recognition of bees by varroa

3.2.1. Olfactometric study
3.2.1.1. Behavioural aspects

The two different olfactometers gave the same results : to sum up, we demonstrated that varroa is attracted by alive workerbees, males and queen, by wax and by cappings of brood, at a mean distance of 0.5 to 1 centimeter.

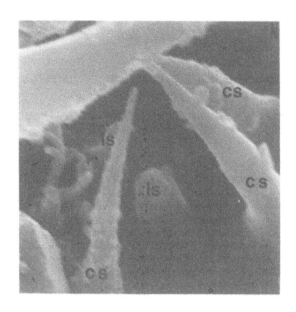

Figure V - Varroa jacobsoni Oudemans 1904 - Sensorial plate of
the first tarsus. x 9500.

Table I : Four-tubes test - Attraction of Varroa by
emergent bees and by 5 days old bees.

Age	Rate of presence on bees (%)	Rate of presence on controls (%)	Number of Varroas	stat. sign.
Older bees	51,3	5,0	310	< 1/1000
Emergent bees	26,2	8,7	80	< 1/1000

Table II : Four-tubes test - Comparison of attractiveness
between older and emergent bees.

Rate of presence on older bees (%)	Rate of presence on emerging bees (%)	Number of Varroa	Statistical Signification
59,7	3,9	70	< 1/1000

3.2.1.2. Chemical analysis

Up to now and with these experimental conditions, none of the extracts has been attractive to varroa.

3.2.2. Study of thermic factors

- Results of the study of varroa's thermopreferendum are shown figure VI. The thermopreferendum depends on the room-temperature, and is between $31.3 \pm 3.3°C$ at 21°C and $34.2 \pm 2.2 °C$ at 34.5°C. The variance analysis gives a significant difference (P<0.01).

- Varroa's age has an effect upon the thermopreferendum, and we can observe a significant difference of 1.6°C between the olders' mean thermopreferendum and the emerging bees'one.

- The four-tubes test used to point out the relative attraction of heat shows that varroa is attracted by heat and is able to discriminate with a 2 degrees difference.

- With the same test, used as a choice test, we demonstrated that varroa is attracted first by heat, second by alive bees, and third by vibrations.

- the percentage of varroas on the bee thorax increases from 1% to 49% when the room-temperature reduces respectively from 34 to 18°C.

3.2.3. Study of vibratory signals

The four-tubes test showed that the vibrations have an effect on the varroa's behaviour : 38% of them go on the tubes in which vibrations are produced, and 13% go on control tubes. This difference is significant (Student test F<0.001).

3.3. Differential attractiveness of workerbees in function of their age

3.3.1. Four-tubes test

- Attraction test shows a more important attraction for older bees (5 days old) than for emerging bees (table I).
- Choice test confirms these results (table II).
- The same results are found with test using aluminium foils.
- Thermic-sounding lines give a 2°C difference between tubes with older bees and tubes with emerging ones.

3.3.2. Passage of varroas from young bees to older ones

- In function of time : Varroas go very quickly from emerging bees to older ones (figure VII). So, after 1 hour experiment, 70% of them are on older bees. The same distribution is obtained after a 24 hours experiment.

- In function of bees'age : Varroas significantly go on older bees rather than on emerging ones (figure VIII). We have observed that, when varroas and emerging bees are placed in a box, half of the varroas leave the young bees and the box. This phenomenom is not observed when they are mixed with 1.5 days old bees.

- In function of room-temperature : There is a room-temperature effect (P<0.05) on varroa's distribution on the two groups of different age (figure IX).

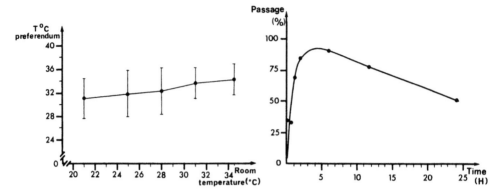

Figure VI - Varroa's temperature preferendum in function of ambiant temperature.

Figure VII - Rate of varroas' passage from emergent bees to 5 days old bees in function of time of experiment.

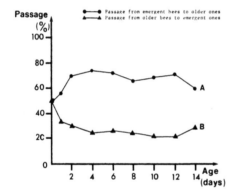

Figure VIII - Rate of Varroa's passage from parazited honeybees to unparazited ones in function of bee ages.

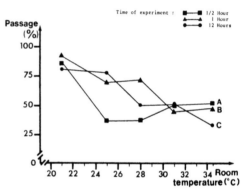

Figure IX - Rate of passage from emergent honeybees to older ones in function of ambiant temperature.

4. Discussion

Varroa uses its first pair of legs as an antenna and the special morphology of the sensory plate placed on them enables us to suppose that it is a specialized system. So, it is interesting to know to which sensory modalities the different types of sensilla correspond.

The olfactometric studies bring us to the following affirmations : One or more molecules are supposed to modify varroa's behaviour so far as varroa is attracted at a distance by wax and cappings. Chemical analyses are in progress.

Results obtained on the influence of heat and of the temperature on varroa's behaviour can be summed up :

                - Varroa is attracted by heat.
                - Varroa is able to make the difference between 2°C
                - Its thermopreferendum varies in function of ambiant

temperature and in function of its age. It corresponds to the heat temperature of brood (34°C for workerbees and 31°C for males (4)) and to the body temperature measured on a bee set apart from the brood (32.4°C on the thorax and 31°C on the abdomen(5)).

    - In these experimental conditions, varroa is attracted first by heat, then by a living bee, then by vibrations produced by a group of 200 bees.

    - The percentage of varroas on bee's thorax (the heating part of bees) increases when ambiant temperature decreases.

    - Varroa's distribution in two different categories of bees (emerging and five days old) is to be understood in function of ambiant temperature.

So, heat and temperature play a fundamental role in the honeybee-varroa relationships.

Varroas don't stay on emerging workerbees. This discrimination blurs out with bees older than one day and a half. Therefore, we can suppose that

    - either emerging bees are repellent,
    - or older bees are more attractive.

However, if we consider:

    1/that varroa is attracted by heat and is able to make the difference between 2°C,

    2/that one emerging bee produces less heat than an older one,

    3/that the repartition of varroa on both groups of bees from different ages varies according to ambiant temperature,

    4/the percentage of varroa on bee's thorax is inversely proportional to ambiant temperature,

then, we can conclude that the different production of heat by bees is a partial explanation for varroa's preference towards older bees. Nevertheless, all this doesn't turn down the hypothesis that older bees produce substances or vibrations that attract varroas, nor the hypothesis that emerging bees generate repellent molecules for varroas.

REFERENCES

1. LE CONTE Y., ARNOLD G., CHAUFFAILLE J., MASSON C., 1984 .-Role of semiochemicals in honeybee. Varroa relationships : preliminary data. 2 nd Meeting of the E.C. varroa Experts'group.

2. RIBBANDS C.R., 1953 .- The behaviour and social life of honeybees. Bee Research Association, London.

3. VET L., VAN LENTEREN J.C., HEYMANS M.,MEELIS E., 1983 .-An airflow olfactometer for measuring olfactory responses of hymenopterous parasitoids and other small insects. Physiological Entomology. 8,97-106.

4. SIMPSON J., 1961 .- Nest climate regulation in honeybee colonies. Science, 133, 1327-1333.

5. HEINRICH B., 1980 .- Mecanisms of body-temperature regulation in honeybees, Apis mellifera. II. Regulation of thoracic temperature at high air temperatures. J. Exp. Biol., 85, 73-87.

The lipids of worker and drone larvae of honeybees (*Apis mellifera* L.)

F.J.Jacobs & F.Vandergeynst
Laboratory for Zoophysiology, State University of Ghent, Ghent, Belgium

Summary

   Lipid extraction of larvae was carried out following FOLCH et al.(1957). The lipids were separated with T.L.C., followed by G.C. (transmethylation of the fatty acids and silylation of the sterols).
   The absence of C16:1 in the classes of free fatty acids and diglycerides for dronelarvae and the presence of C18:2 in the classes of triglycerides and phospholipids for dronelarvae suggests different metabolic pathways of lipids in drone- and workerlarvae.
   The increasing amount of stigmasterol in dronelarvae(L5) seems to be very important.

1. Introduction

   Female mites of Varroa jacobsoni prefer dronecells in order to deposite their eggs. In relation to this fact, it is important to compare the chemical substances in last-instar workerlarvae and last-instar dronelarvae. In relation to Nosematosis, VAN DER VORST et al. (1981 and 1983) determined already the lipid- and sterolcomposition of workerlarvae. So it became obvious to compare these results with lipid analyses of L5 dronelarvae.
   The ultimate goal is to find out if lipids and sterols have a control function in relation to the preference of the mite for dronelarvae. But these analyses are also important in relation to the physiology of the nutrition of the mite.

2. Materials and methods

   Larvae were token from one bee colony (A. m. carnica F1) on days 1, 2, 3, 4 and 5 for workerlarvae and on day 3, 4, 5 and 7 for dronelarvae. Following the method of REMBOLD et al. (1980), the age of each larva was determined by weighing. The dry weight was determined after homogenising and drying overnight (110 °C).
   Lipid extraction was carried out, following FOLCH et al. (1957). The lipid classes were separated by thin layer chromatography (Silica gel G 60, 0.5 mm thick, using a solvent mixture of n-hexane, diëthylether, formic acid 80:20:2). The fatty acids were transmethylated, quantitatively and qualitatively analysed, according to CHRISTIE et al. (1970).

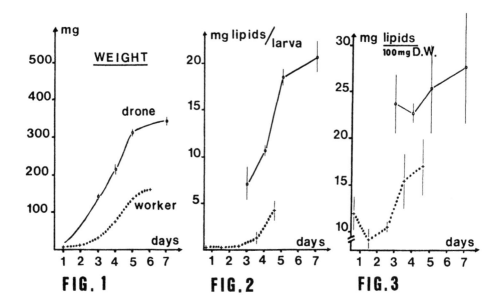

FIG. 1   FIG.2   FIG.3

Pentadecanoate was used as internal standard in the gaschromatographical analyses of fatty acids. A fused silica open tubular column (25 m x 0.25 mm int. diam.) was used with SUPEROX FABP as station ar phase (0.2 µm). The injector and detector (F.I.D) chamber was placed at 220 °C. The temperature of the column was programmed from 115 °C, during 10 min, increased with 3 °C each minute, untill 196 °C was reached. Hydrogen was used as carrier gas (flow rate of 1.3 ml/min; in the detector, a make-up flow of 25 ml/min was used).

Cholestanol was used as internal standard for sterol-analyses. After sililation with 20 µl of tri-sil/BSA formula P1 for 10 min at 70 °C, separation was carried out on a SP 52 WCOT column (50 m and 0.5 mm int. diam.). Hydrogen was used as carrier gas (flow rate : 4 ml/min. The temperature of the coilumn was programmed from 150 °C to 260 °C at 6 °C/min. The temperature of the injector and detector was held constantly on 230 °C.

3. Results

In fig. 1 the larval weight is plotted against the age of the larvae. The fresh weight is increasing exponentially and during the development of the larvae the ratio lipid weight/ dry weight increases. And this ratio is reaching for dronelarvae twice the level of workerlarvae (fig. 3).

The amount of detected fatty acids in the class of phospholipids is given in table nr. 1. Linoleic acid (C18:2) is only present in the phospholipids of dronelarvae. Oleic acid (C18:1) is the most abundant fatty acid, both for worker- and dronelarvae.

In table 2, the fatty acids in the diglycerides are presented. The concentration of palmitoleic acid (C16:1) is

Table nr. 1   µg/ larva of phospholipids on day :

		3	4	5	3	4	5	7
		W O R K E R			D R O N E			
C16	X	3	34	88	169	118	326	103
	N	3	10	5	5	4	3	5
	SD	1	12	30	14	77	52	12
C16:1	X	2	6	8	12	7	10	ND
	N	3	10	5	5	4	3	5
	SD	1	3	3	2	4	8	
C18	X	3	29	79	161	50	43	167
	N	2	10	5	5	4	3	5
	SD	1	9	41	40	21	46	50
C18:1	X	25	189	483	481	349	1473	406
	N	3	10	5	5	4	3	.5
	SD	11	54	218	38	141	450	73
C18:2	X				14	11	28	12
	N				5	4	3	5
	SD				2	6	25	1
C18:3	X	1	6	12	20	16	67	39
	N	3	10	5	5	4	3	5
	SD	1	4	4	2	9	35	3

Table nr. 2.   µg/ larva of diglycerides :

		3	4	5	3	4	5	7
C16	X	1	4	17	30	18	62	53
	N	1	6	4	3	5	3	5
	SD		2	4	5	14	13	13
C16:1	X	1	1	1				
	N	1	6	4				
	SD		1	1				
C18	X	1	1	5	10	7	11	12
	N	1	6	4	3	5	3	5
	SD		1	1	3	2	2	3
C18:1	X	2	7	25	52	23	68	52
	N	1	6	4	3	5	3	5
	SD		2	4	4	5	11	10
C18:2	X	ND	ND	1	2	2	2	ND
	N			4	3	5	3	5
	SD			1	1	1	1	
C18:3	X	ND	1	2	2	2	2	ND
	N		6	4	3	5	3	
	SD		1	1	1	1	1	

X = the mean ; N = number ; SD = standard deviation
ND = present, but in a so small amount that quanti-
    sation is impossible.

Table nr. 3   µg/ larva of triglycerides on day :

		3	4	5	3	4	5	7
			W O R K E R			D R O N E		
C14	X	ND	12	40	147	221	443	307
	N		9	7	4	5	3	5
	SD		6	22	13	44	22	78
C16	X	4	142	553	1901	2988	5308	3945
	N	3	9	7	4	5	3	5
	SD	4	82	319	206	516	397	260
C16:1	X	1	6	12	25	22	25	38
	N	3	9	7	4	5	3	5
	SD	1	2	7	14	5	12	14
C18	X	1	46	170	927	1928	3683	2644
	N	3	9	7	4	5	3	5
	SD	1	28	98	252	696	620	260
C18:1	X	5	140	518	443	571	726	1002
	N	3	9	7	4	5	3	5
	SD	5	77	289	134	472	95	200
C18:2	X				ND	11	49	25
	N					5	3	5
	SD					9	35	9
C18:3	X	ND	ND	9	ND	18	15	34
	N			7	4	5	3	5
	SD			7		12	3	13

Table nr. 4.   µg/ larva of free fatty acids:

		3	4	5	3	4	5	7
C16	X	4	6	12	44	41	97	18
	N	3	2	4	5	5	3	6
	SD	1	2	4	6	18	20	6
C16:1	X	1	1	1				
	N	3	2	4				
	SD	1	1	1				
C18	X	2	2	5	13	12	22	6
	N	3	2	4	5	5	3	6
	SD	1	1	1	4	4	4	1
C18:1	X	9	12	22	36	37	83	20
	N	3	2	4	5	5	3	6
	SD	5	2	5	6	7	12	6
C18:2	X	ND	ND	1	5	7	20	ND
	N			4	5	5	3	6
	SD			1	2	2	7	
C18:3	X	ND	1	2	4	7	39	ND
	N		2	4	5	5	3	6
	SD		1	1	3	2	10	

Symbols : see table 1 and 2.

Table nr. 5   µg/ larva of free sterols on day :

		\ 3	4	5	3	4	5	7
		W O R K E R			D R O N E			
Choleste-	X	0.13	0.29	1.01		ND	3.45	1.60
rol	N	6	11	7			6	6
	SD	0.08	0.17	0.29			0.66	0.62
24-methy-	X	2.39	9.69	29.24	58.86	74.97	92.72	43.45
leenchol.	N	6	11	7	7	6	6	6
	SD	0.70	4.35	6.26	6.27	6.94	15.14	4.40
Stigma-	X	0.34	1.39	4.30	12.16	16.92	30.78	49.71
sterol	N	6	11	7	7	6	6	6
	SD	0.10	0.58	0.93	0.89	1.22	5.99	9.24
Avena-	X	0.32	1.77	6.03	7.02	9.39	18.71	23.06
sterol	N	6	11	7	7	6	6	6
	SD	0.14	0.91	1.33	0.99	1.64	3.24	1.73

decreasing from day 1 to day 5 for workerlarvae. In dronelarvae C16:1 is absent. Oleic acid again is the most abundant fatty acid.

The results for the class of triglycerides are given in table 3. Comparably with the phospholipids is linoleic acid not detectable in triglycerides of workerlarvae. The most abundant fatty acid is palmitic acid (C16). Much more C18 (saturated) is present in the triglycerides of drones and less C18:1 (unsaturated).

The amount of free fatty acids is given in table 4. C16:1 is absent in drone-, but present in workerlarvae. The most important fatty acids are C16 and C18:1.

From the data in table nr. 5 can be concluded that the most striking phenomenon in the free sterols can be stated in the increase of stigmasterol in dronelarvae: from 15.6 % to 42.2 %.

4. Discussion

Larval tissue of drones contains the double amount of lipids, compared with workerlarvae. This difference is particularly caused by an increased synthesis of triglycerides. Triglycerides serves as energy source during pupation, so it seems to be logical in relation to the pupation period, which is longer for dronelarvae.

By VAN DER VORST et al. (1981), the hypothesis was put forward that the rate of lipid synthesis increases during larval development. This phenomenon is established again for dronelarvae. Moreover, the absence of C16:1 in the classes of free fatty acids and diglycerides for dronelarvae and the presence of C18:2 in the classes of triglycerides and phospholipids for dronelarvae suggests a different metabolic pathway for the synthesis of lipids in honeybee larvae. Concerning the free sterols, the increasing amount of stigmasterol seems to be very important.

<u>References</u>

CHRISTIE,   W.W.;   NOBLE,   R.C.   and   MOORE,   J.H.   (1970)
            Analyst 95 : 940 - 944

FOLCH, J.; LEES, M. and STANLEY, S. (1957)
            J. biol. Chem. 226 : 397 - 409

REMBOLD, H.; KREMER, J.P. and ULRICH, M.G. (1980)
            Apidologie 11 (1) : 29 - 38

VAN DER VORST, E.;   DE RYCKE, P.H.;   MATTIJS, J. and JACOBS
            F.J. (1981)   28 Congres Apimondia, Bukarest :
            294 - 298.

VAN DER VORST, E.;   MATTIJS, J.; JACOBS, F.J. and DE RYCKE
            P.H.(1983)   J. of Apicult. Res. 22 (1): 3 - 8

Reproduction of 'standard' Varroa mites in relation to their preceding stay in adult bees of different age and function

J. Van Esch & J. Beetsma
Department of Entomology, Wageningen Agricultural University, Wageningen, Netherlands

Summary

The rate of reproduction of adult females, after introduction into recently sealed worker brood cells, is affected by their origin and appearance. For the present study standardization of the mites seemed
essential. As standard mites only adult females known to have reproduced at least one time were used. These mites were collected from worker brood cells 10 days after cell sealing immediately after removal
of the comb from the colony. Subsequently the mites were kept in an
incubator for one week on different categories of caged bees:
1. newly emerged bees, 2. "nurse bees" 15 days of age, collected from
a colony containing all brood stages, 3. "winter bees" 15 days of age,
collected from a colony in which the queen was confined to one comb
and from which every third day all eggs were removed, 4. foragers
30 days of age.
After introducing the mites into newly sealed worker brood cells it
appeared that the percentages of reproducing mites and the average
numbers of offspring per mite did not differ significantly between
the treatments.

1. Introduction

In addition to the finding that Varroa mites are not randomly distributed on worker bees of different ages (7), data concerning their subsequent reproduction in worker brood cells could lead to the development of
a method to control the Varroa mite in honeybee colonies during the active
season without using acaricides.

In preceding studies considerable differences were found in the rate
of reproduction, when adult females were used immediately after their collection from adult bees or brood cells, or after storage of bees or brood
in an incubator (pers.comm. DE RUIJTER).

When collecting adult females from Apis m. intermissa worker bees,
BEETSMA and ZONNEVELD found that these females demonstrated highly varying
rates of reproduction. The mites clearly differed in the fact that they
were either swollen or not swollen. In the swollen mites the dorsal and
ventral shields were separated. Both the percentage of swollen females
that produced adult offspring and their average number of offspring per reproducing female, within 10 days after introduction into recently sealed
worker brood cells, were higher than in non-swollen females (1).

The condition of the adult and immature host and the stay on caged bees in an incubator apparently affect the rate of reproduction of the Varroa mite. Using worker bees and brood, from a colony kept in a flight-room, and adult females, collected from bees of a colony overwintering in the field, HÄNEL found that only 4.5 % of the adult females produced offspring when introduced into recently sealed worker brood cells subsequent to collection. The percentage increased to 10.8 when the adult females had been kept on caged adult bees in an incubator during 7.8 days (range 1-31 days) before introduction into worker brood cells (4).

In the present study it is tried to reduce the variability in the rate of reproduction by standardization of the mites used. Old females, collected from worker brood cells 10 days after cell sealing, that had produced offspring seemed to be the most obvious choice.

Age groups of adult worker bees can readily be obtained by introducing marked newly emerged bees into the colony. However, functional groups of worker bees can only be collected from observation hives (2). Standardization of groups of worker bees according to their physiological condition is complicated by the fact that few data are available concerning both the physiological changes in the worker bee and the nutritional requirements of the Varroa mite.

We assumed that worker bees 15 days of age, collected from a colony containing all stages of brood, functioned as nurse bees. The age of nurse bees feeding worker larvae in an observation hive varies according to LINDAUER between 1 and 28 days (8).However, when BROUWERS measured the protein synthesis of isolated hypopharyngeal glands from bees 0-17 days of age, collected from a colony during summer, the highest percentage (33 %) of bees with glands that had the same stage of development and demonstrated a similar rate of activity as found in those from known nurse bees, was found in the bees 8-17 days of age (2).

We assumed that worker bees 15 days of age, collected from a colony kept broodless for at least 3 weeks, had become winter bees. MAURIZIO found that winter bees can be obtained during summer by restricting the queen to a queen cage (9). Worker bees that are not allowed to nurse brood are longlived and demonstrate a well developed fat body.

Although DE RUIJTER and PAPPAS demonstrated that the rate of reproduction of adult females is affected by a preceding stay on adult worker bees (10), few data are available concerning the relation between age and function of the adult host and subsequent reproduction of adult females in worker brood cells.

After a stay of 3-5 days on caged bees 2, 5 and 12 days of age in an incubator, HÄNEL found that respectively 0, 22 and 50 % of the adult females produced offspring in worker brood. These differential percentages of reproducing adult females were discussed in relation to the juvenile hormone titre in the hemolymph of worker bees of different ages.

The aim of the present study is to compare the rate of reproduction of "standard" mites in worker brood cells after a stay on adult worker bees of different age and function.

2. Materials and methods

The experiments were carried out in Wageningen from June until the beginning of September 1985 using a hybrid strain of Apis m. mellifera L. The level of infestation with Varroa mites of the colonies was very low. Adult females were collected from sealed worker brood cells, 10 days after cell sealing, from heavily infested colonies kept on the campus of the Technical University Twente in Enschede. Only females that had reproduced were used. In case two adult females occurred in one brood cell, these females were excluded from the experiment.

Standard mites were kept on:
1. Newly emerged bees,
2. Marked bees 15 days of age ("nurse bees") and
3. Marked bees 30 days of age (foragers), collected from a colony containing a queen and all stages of brood.
4. Marked bees 15 days of age ("winter bees"), collected from a colony in which the queen had repeatedly been confined to one comb. The combs containing eggs were removed every third day.

Only "nurse bees" received a piece of comb containing open brood. All categories of caged bees were placed in an incubator at 33°C. and provided with water and a pollen-sugar mixture (3). After 7 days the mites were collected from the bees and introduced into recently (0-4 h) sealed worker brood cells. The comb was reintroduced into the colony. The presence and the stage of development of offspring were observed 10 days after introduction of the mites into the brood cells.

The offspring of an introduced adult female could not be noted, when the brood cell was empty, when the female had disappeared or died, when the bees had fastened the mite in the cell capping or when two or three adult females were found in one brood cell.

Differences in the average numbers of offspring per adult female between treatments were studied using the U-test of WILCOXON, MANN and WHITNEY (11).

TABLE I. Average numbers of offspring (+S.D.) of standard <u>Varroa</u> females, kept on different categories of worker bees in an incubator for 7 days, produced within 10 days after introduction into sealed worker brood cells. "winter" and "nurse": presumptive winter- and nurse bees. E/L: egg-larva; PROTO: protonymphs; DEUTO: deutonymphs; A: active-; P: passive stage; M: male; F: female.

CATEGORY & AGE OF BEES IN DAYS	NO.OF MITES	E/L M+F	PROTO A M+F	PROTO P M+F	DEUTO A M	DEUTO A F	DEUTO P M	DEUTO P F	ADULT M	ADULT F	SUM
"winter" 15	18	0.3 +0.8	0.2 +0.4	0.5 +0.5	0.1 +0.2	0.2 +0.4	0.1 +0.2	1.2 +0.8	0.6 +0.5	0.4 +0.5	3.6 +1.2
newly emerged	0 21	0.3 +0.6	0.1 +0.3	0.4 +0.7	0.0 +0.0	0.3 +0.5	0.0 +0.0	0.9 +0.8	0.5 +0.5	0.6 +0.6	3.2 +1.8
"nurse" 15	13	0.2 +0.4	0.1 +0.3	0.5 +0.5	0.0 +0.0	0.3 +0.5	0.2 +0.4	1.2 +0.9	0.6 +0.5	0.8 +0.6	3.8 +1.5
forager 30	9	0.2 +0.7	0.1 +0.3	0.1 +0.3	0.0 +0.0	0.4 +0.5	0.0 +0.0	1.1 +0.9	0.9 +0.3	1.0 +0.5	3.9 +1.4

3. Results

Small differences in the percentage of reproducing females were found between mites kept on different categories of bees (TABLE II). No significant differences could be demonstrated in the average number of offspring (living and dead) per adult female, within 10 days after introducing them into worker brood cells, between the treatments (TABLE I).Dead individuals were found in the offspring of adult females kept on different categories of worker bees. Respectively 10.9 %, 14.7 %, 4.1 % and 2.9 % of the offspring of adult females kept on "winter bees", newly emerged bees, "nurse bees" and foragers had died. When considering only the living individuals, the average numbers of offspring (+S.D.), 3.2+1.7, 2.8+1.6, 3.6+1.6 and 3.8+1.5 respectively, did not differ significantly either.

The percentage of mothers producing 1-2 adult daughters, within 10 days after introducing them into worker brood cells, increased from winter bees to foragers. The average number of adult daughters per female was significantly ($P < 0.05$) higher in females kept on foragers than in those kept on winter bees (TABLE II). No significant differences could be demonstrated between these two and the other treatments.

However, the difference found is only due to a differential stage of development of the offspring. When using data of IFANTIDIS concerning the duration of the developmental stages of the Varroa mite (5), the number of adult daughters present at the 12th day after introduction of the mothers into worker brood cells can be estimated. When assuming that maximal one

TABLE II. Observed and estimated success of reproduction (+S.D.) of standard Varroa females, kept on different categories of worker bees in an incubator for 7 days, within 10 days after introduction into sealed worker brood cells

Abbreviations: see Table I.

CATEGORY OF BEES	% OF MITES WITH OFFSPRING	% OF MITES WITH 1-2 ADULT DAUGHTERS	AVERAGE NUMBER OF ADULT DAUGHTERS	% OF MITES WITH 1-2 ADULT DAUGHTERS	AVERAGE NUMBER OF ADULT DAUGHTERS
		OBSERVED 10 DAYS AFTER INTRODUCTION INTO WORKER BROOD CELL		ESTIMATED 12 DAYS AFTER INTRODUCTION INTO WORKER BROOD CELL	
"winter"	100	44	0.44 +0.51	78	1.44 +0.86
newly emerged	91	57	0.62 +0.59	71	1.29 +0.90
"nurse"	92	69	0.77 +0.60	77	1.39 +0.87
forager	100	89	1.00 +0.50	89	1.56 +0.73

passive female deutonymph developed into an adult daughter within two days, the percentages of mothers with 1-2 adult daughters demonstrate smaller differences and the average numbers of adult daughters per adult female are not significantly different (TABLE II).

4. Discussion

The occurrence of non-reproducing adult females may be partly due to damage caused by their introduction into brood cells. BEETSMA and ZONNEVELD found a similar percentage of reproducing females (100 %) after natural infestation of A. m. intermissa worker brood (1), however, SCHULZ and IFANTIDIS found lower percentages using A. m. carnica (84 %) (12) and A. m. cecropia (53 %) (5.6) under the same conditions.

The percentages of adult females producing 1-2 daughters and the average numbers of adult daughters per female, within 10 days after introducing them into worker brood cells, demonstrated in the present study, are mostly higher than after natural infestation of A. m. intermissa worker brood cells, 52 % and 0.55 respectively (1).

Our estimated percentages of adult females producing 1-2 daughters and the average numbers of adult daughters per female in worker brood cells are in agreement with the estimates of SCHULZ using A. m. carnica, 73 % and 1.3 respectively (12), but higher than the observations of IFANTIDIS using A. m. cecropia, 53 % and 0.71 respectively (6).

Our results are not in agreement with the finding of HÄNEL that the percentage of reproducing adult females increases with the age of the worker bees on which they were kept for 3-5 days (4). This differential result could be due to our selection of standard mites.

In the present study the adult females were kept on worker bees for 7 days. This period is comparable to the stay of marked adult females on worker bees before their entering into brood cells for their second reproduction cycle as observed by SCHULZ (12).

The discrepancy between the earlier results, when the mites were kept on adult worker bees for 3-5 days (4), and those of the present study could also be due to their prolonged stay on caged bees in an incubator. The physiological conditions of these bees might have changed into a uniform "incubator bee" condition.

5. Acknowledgements

We want to express our gratitude to the Board and Mr.H.J.Lutke Holzik of the Technical University Twente for all facilities placed at our disposal to carry out our studies with Varroa infested colonies at the campus in Enschede.

We thank our colleague F.P.Wiegers for his cooperation and fruitful discussions.

References

1. BEETSMA, J.and ZONNEVELD, K. (in preparation) Preliminary studies of the reproduction of the Varroa mite in relation to its stay on adult and immature honeybee workers
2. BROUWERS, E.V.M. (1982) Measurement of hypopharyngeal gland activity in the honeybee. J.apic.Res. 21(4):193-198
3. GROOT, A.P.DE (1953) Protein and amino acid requirements of the honeybee (Apis mellifica L.). Physiol.Comp. et Oecol.3:197-285
4. HÄNEL, H.(1985) Das Juvenilhormon der Honigbiene als ein Auslöser für die Reproduktion des Parasiten Varroa jacobsoni Oud. unter Berücksichtigung der Morphologie des Reproduktionssystems der Milbe. Dissertation Johann Wolfgang Goethe Universität, Frankfurt am Main

5. IFANTIDIS, M.D.(1983) Ontogenesis of the mite Varroa jacobsoni in worker and drone honeybee brood cells. J.apic.Res. 22(3): 200-206
6. IFANTIDIS, M.D.(1984) Parameters of the population dynamics of the Varroa mite on honeybees. J.apic.Res. 23(4): 227-233
7. KRAUS, B.(1985) In: DRESCHER,W.,MADEL,G. Bericht von der Tagung der Arbeitsgemeinschaft der Institute für Bienenforschung e.V., in Bonn vom 12. bis 14. März 1985. ADIZ 10: 313-317
8. LINDAUER, M.(1952) Ein Beitrag zur Frage der Arbeitsteilung im Bienen-staat. Z.für vergl.Physiol. 34: 299-345
9. MAURIZIO, A.(1950) Untersuchungen über den Einfluss der Pollennah-rung und Brutpflege auf die Lebensdauer und den physiologischen Zu-stand von Bienen. Schweiz.Bienenztg. 73: 58-64
10. RUIJTER, A.DE and PAPPAS, N.(1983) Karyotype and sex determination of Varroa jacobsoni Oud. Proceed.Meeting of the EC Experts' Group, 7-9 February, Wageningen: 41-44
11. SACHS, L.(1974) Angewandte Statistik. Springer-Verlag Berlin, Heidel-berg New York
12. SCHULZ, A.E.(1984) Reproduktion und Populationsentwicklung der parasi-tischen Milbe Varroa jacobsoni Oud. in Abhängigkeit vom Brutzy-klus ihres Wirtes Apis mellifera L. Apidologie 15(4): 401-420

Reproduction of Varroa mites during successive brood cycles of the honeybee

A.De Ruijter
Research Centre for Insect Pollination and Beekeeping 'Ambrosiushoeve', Hilvarenbeek, Netherlands

Summary

Newly capped worker brood cells were numbered on a sheet of transparent plastic, temporarily attached to the top bar using two thumbtacks. Into each cell a female Varroa mite was introduced. After ten days the cells were opened and the content studied. Those females still present and alive were introduced into newly capped brood cells and so on.
Varroa mites are capable of reproducing up to seven times this way. The maximum number of eggs layed was 30 eggs per female. Females that produce male offspring only, keep doing so in subsequent brood cycles. Though in contact with adult males several times, no successful mating had occurred. Probably only young females mate successfully.

1. Introduction

During the last few years substantial progress has been made in chemical control of Varroa jacobsoni (3, 4, 5). However, most scientists working on Varroa control agree on the fact, that more knowledge has to be gathered on the biology of the mite,in order to develop biotechnical control methods (1). In this respect the reproduction of Varroa is of utmost importance.

Although sampling of Varroa infected brood can give information concerning reproduction, more information can be gathered when mites are reared inside the colony, using the method developed by Ifantidis (2).

The objectives of this study were to establish the maximum number of successive reproduction cycles and the maximum number of eggs layed by a single female.

2. Materials and Methods

Worker brood cells containing old larvae are marked on a sheet of transparent plastic, temporarily attached to the top bar, using two thumbtacks. The comb is put back onto the colony and a few hours later the cells that are capped are numbered. From capped brood cells adult female mites,both"mothers" and "daughters", are collected. Into each newly capped cell one of these female Varroa mites is introduced by opening the cell carefully, letting the mite in and closing the cell thoroughly again.

After 10 days the cells are opened to examine the content. Those females still present and alive are introduced into newly capped brood cells and so on.

3. Results and Discussion

In 315 cases the female mite was regained alive, 250 times with offspring, 65 times without offspring (20.7%). If there is no offspring, this does not always mean that the mite is infertile. Twenty one mites interrupted egg laying for one period or more and resumed egg laying in the next period. This may be due

63

to the lack of a stimulus, as a result of the haemolymph composition of the pupa. Six mites did not lay eggs at all. The percentage of mites without offspring may be an important criterion for selecting Varroa resistant bees. It should be kept in mind however, that the first time mites are introduced, this percentage may be slightly higher than later on, because very young mites don't lay eggs during the first period

Figure 1 shows the reproduction in 8 successive periods of 10 days. Offspring was found in 7 periods. Only a few mites were found to be dead. Most of the mites disappear because the bees refuse the cells we have opened and not closed completely again. The bees uncap these cells and clear them out.

The frequency of the number of succesfull reproductive cycles is shown in fig. 2. Zero or one successful cycle are the most frequent, 2, 3, 4, 5 or 6 cycles occur also and the maximum is 7 cycles with offspring. If we exclude the mites that were lost, due to the method, and only look at the mites that were found dead or were regained alive, 5 or 6 successful cycles are the most frequent (fig. 3).

To find out if there is a decrease in the number of eggs layed in successive periods, we selected those mites that had offspring in 4 periods or more. When offspring of both sexes occurs (fig. 4), the number of eggs per period is about four, with a slight decrease until the sixth period and then drops to two eggs. In those cases where females were unmated and had male offspring only, the average number of eggs is about 2 per period. The total number of eggs layed per individual female increases with the number of successful periods (fig. 5). The maximum number of eggs layed by unmated females is much smaller. Females that produce male offspring only, keep doing so in subsequent brood cycles. Though in contact with adult males several times, no successful mating had occurred. Probably only young females mate successfully.

4. Conclusion

Varroa mites are able to reproduce up to seven times and are capable to produce up to 30 eggs. Contact with adult bees is not indispensable for reproduction. In spite of our manipulating these mites, some of them lived over 2.5 months in summer.

Whether this maximum reproduction is reached in practice depends on different factors inside the bee colony, of which we only have limited knowledge.

REFERENCES

1. Arbeitsgemeinschaft der Institute für Bienenforschung (1986). Empfehlungen zur praxisorientierten Varroatose-Forschung. Allg. Deutsche Imkerzeitung 20 (9), 288
2. Ifantidis, M.D.(1980). Ontogenesis of the mite Varroa jacobsoni O. in the worker and drone brood cell of the honey bee Apis mellifera cecropia. J. Apic. Res. 22 (3), 200-206.
3. Ritter, W.; Perschil, F. and Czarnecki, J.M. (1983). Treatment of bee colonies with isopropyl-4,4-dibromo-benzilate against Varroa disease and Acarine disease. Zbl. Vet. Med. B, 30, 266-273
4. Ritter, W. (1986). Die Varroatose der Honigbiene, Apis mellifera, und ihre Bekämpfung mit Perizin. Veterinär-Medizinische Nachrichten, 1986 (1), 3-16
5. Ruijter, A. de; Eijnde, J. van den (1986). Field experiment to determine the effects of Perizin on Varroa mites and on development of treated colonies. Veterinary Medical Review 2/86, in press.

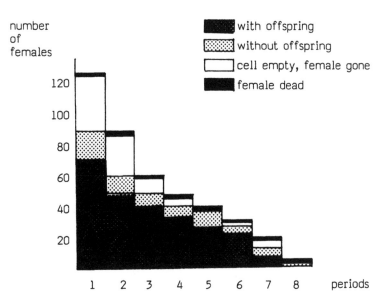

Fig. 1.   Reproduction during 8 successive periods of 10 days each.

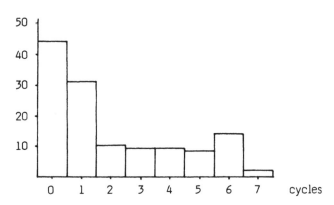

Fig. 2.   Frequency of the number of successful reproduction cycles.
All females.

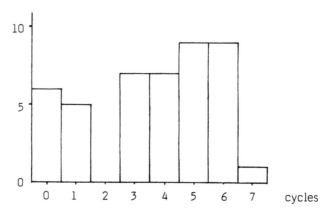

Fig. 3.   Frequency of the number of successful reproduction cycles.
Females that were lost excepted.

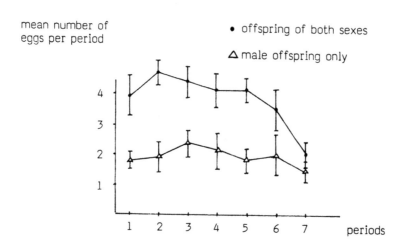

Fig. 4.   Mean number of eggs per period from females with offspring
in 4 periods or more.

Fig. 5.   Total number of eggs per female (maximum, mean and minimum).

Individual differences in *Varroa jacobsoni* of preference for drone larvae to worker bee larvae

C.Otten & S.Fuchs

Institut für Bienenkunde (Polytechnische Gesellschaft), Fachbereich Biologie, J.W.Goethe Universität Frankfurt am Main, Oberursel, FR Germany

SUMMARY

By continuously removing either capped drone cells (worker line) or capped worker cells (drone line) from two sets of colonies respectively, types of Varroa could be separated which differed in their preference for drone larvae in a choice test. Drone-line Varroa spent longer times on drone larvae than worker-line Varroa, whereas both types spent about equal times on drone larvae.

1 Introduction

The preference of Varroa for drone cells in comparison to worker cells is based on stimuli coming from the larvae (1). It was found to be 8.6:1 in colonies of A. mellifera carnica (2) and 4:1 in A. cerana. In Uruguay, drone cell preference increased substantially within two years (3). Also these ratios may reflect characteristics of the bee larvae, the possibility is considered that the degree of drone cell preference is an adaptive trait of Varroa set by natural selektion. As a prerequisit, individual variation of drone cell preference should be found in Varroa populations. The current study aimed to prove the existence of Varroa types with high or low drone preference, which may be separated by a process of selection.

2 Methods

In 15 intervals of 7 d, (28.5.-3.9.), one drone comb and one worker comb with brood no older than 7 d since oviposition were placed into colonies. After capping, drone combs were removed in three colonies (worker line; totals 63000 worker cells and 12000 drone cells, 74% of which were removed). In three other colonies, worker combs were removed (drone line; totals 7100 drone cells and 39000 worker cells, 87% of which were removed). By this, almost all Varroa mites were eliminated which had chosen to enter drone cells or worker cells respectively. Colonies were maintained by adding Varroa-free capped brood combs.

Preference tests were carried out during the last 3 weeks of selektion using Varroa taken from the abdominal sternites of worker bees. In the middle of testing arenas (diameter 2.5 cm) with wax bottom, a drone larva and a worker larva from newly capped cells were placed in close contact. This allowed the mites to change between the larvae without touching ground. The time was registrated which the mites spent on the drone larva or worker larva. Each Varroa was tested in 12 trials of 10 min duration, carried out in successive arenas. At the start of a trial, a Varroa was placed on the drone larva or the worker larva. Starting positions alternated in successive trials. Mites of each selektion line were tested in pairs starting at the same larvae (Fig. 1).

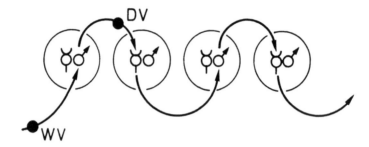

Fig. 1: Testing schedule for pairs of Varroa mites in successive arenas, each containig a drone larva and a worker larva. WV=worker-line varroa, DV=drone-line Varroa. The next pair would start at the worker larva.

3 Results

During 162 trials of 10 min duration, the mean time spent on drone larvae was 5.1 min in worker-line Varroa and 6.4 in drone-line Varroa (p‹0.0005, Wilcoxon paired rank test). In 15 of the 18 pairs tested, drone-line Varroa spent longer times on drone larvae than worker line Varroa (1 pair: no difference, 2 pairs: shorter times; p‹0.01, Sign test).

Within trials, the proportion of time spent on a larva depended on the starting position. If first put on worker larvae, drone-line Varroa spent shorter times on these than worker-line Varroa (Fig.2; means: 5.4 min and 8.1 min respectively, p‹0.02). If put on drone larvae first, mean times differed only slightly (1.2 min and 1.6 min repectively)

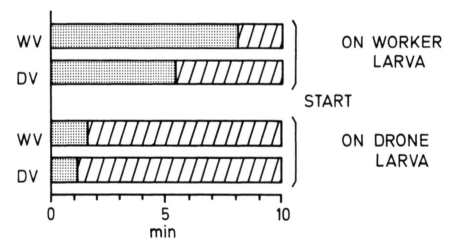

Fig. 2: Mean times spent on worker larvae or drone larva during trials in relation to the starting position. Stippled areas: Times on worker larvae. Hatched areas: Times on drone larvae. WV=worker-line Varroa, DV=drone-line Varroa.

4 Discussion

The highly significant difference between Varroa from both lines in regard to the total number of trials shows very clearly, that individual differences of drone larva preference do exist. In addition, Varroa of high drone preference were significantly more frequent in the drone -selection colonies. This shows, that Varroa types can be efficiently selected by manipulating the survival/reproduction chances in the brood cell types.

The main difference in the trials was, that Varroa from the drone-selection line spent much shorter times on worker larvae than those of the worker-selection line if placed on them as a starting position. In contrast, an only slight difference occurred, if the starting position was a drone larva. Apparently, the discrimination of the larva types is not working two ways by identifying both, drone larvae and worker larvae. This indicates a mechanism of the kind of a graded characteristic of the larvae with a higher level in the drone larvae, to which drone-line Varroa react with a higher behavioral threshold for entering or staying.

REFERENCES

1. ROSENKRANZ, P. et al. (1984): Optimal host selektion by reproductive female Varroa jacobsoni. In: W. ENGELS (ed.): Advance in Invertebrate Reproduction 3, 628. Elsevier Amsterdam, N.Y., Oxford

2. SCHULZ, A.E. (1984): Reproduktion und Populationsentwicklung der parasitischen Milbe Varroa jacobsoni Oud. in Abhängigkeit vom Brutzyclus ihres Wirtes Apis mellifera L. Apidologie 15 (4),401-420

3. KOENIGER, N. et al. (1983): Observations on mites of the Asian honeybee species (Apis cerana, Apis dorsata, Apis florea). Apidologie 14 (3), 205-224

4. RUTTNER,F. et al. (1984): Beobachtung über eine mögliche Anpassung von Varroa jacobsoni an Apis mellifera L. in Uruguay. Apidologie 15 (1), 43-62

Supported by the Deutsche Forschungsgemeinschaft (Ko 400/7-1)

The distribution of *Varroa jacobsoni* on honeybee brood combs and within brood cells as a consequence of fluctuation infestation rates

S.Fuchs

Institut für Bienenkunde (Polytechnische Gesellschaft), Fachbereich Biologie, J.W.Goethe Universität Frankfurt am Main, Oberursel, FR Germany

SUMMARY

The numbers of cells of a brood comb entered by 0,1,2,... etc. Varroa mites differ from a random distribution (poisson-distribution) by higher numbers of cells with few or many mites and lower numbers containing medium numbers. Brood comb areas containing capped brood of different ages varied widely in their infection rates. Model calculations showed, that this variance may account for the observed deviations in the cell infestation frequencies. Additionally, they indicated, that variation of infection rates is rather caused by substantial fluctuations of the numbers of infective mites than of variations in the numbers of available cells.

1 Introduction

The distribution of the infestation of brood combs or cells gives some insight into the complexity of the factors influencing the transition of Varroa from bees into cells. Simple and constant conditions, as constant numbers of Varroa seeking to enter cells, constant cell availability, equal attractivity of the cells and no interactions between mites will lead to distributions resembling random patterns. More complex conditions will cause deviating distributions. Moreover, the actual distributions may help to understand the factors governing the dynamics of brood infestation.

2 Results

Fig. 3 gives examples of a worker comb and a drone comb, where the infestation of each cell had been recorded. The numbers of cells with 0,1,2,... mites deviate significantly from poisson distributions calculated from the average comb infestation (Varroa/cell, V/C). Cells with medium numbers of mites are less frequent in favour of cells with no or few mites and of cells with high numbers. This typical feature is more pronounced if infestation rates are high and was consistently found in about 100 samples studied (1).

The infestion of brood combs differs substantially between brood comb areas. Without apparent regularity, areas of high density and areas of low density are scattered over the brood comb and can be proven statistically as non-random effects. Fig. 2 gives an example of the distribution of Varroa infestation of a worker brood comb.

Areas of higher or lower cell infestation rate often coincide with differences in the developmental stage of the brood. Fig. 3 gives the infestation rates of three developmental stages of bee brood for three capped worker brood combs. Infestation rates differed irregularly and significantly between the age groups of worker brood.

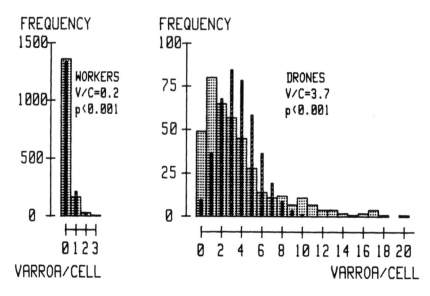

Fig.1: Distribution of cell infestation frequencies (stippled columns) in a worker comb (left) and a drone comb (right). Hatched columns: Poisson distributions. Statistical significance of difference between curves: Chi^2-tests

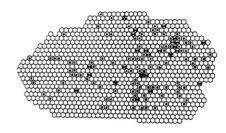

Fig. 2: Ihe distribution of Varroa infestation of a capped worker brood comb. Each dot represents one mite

Discussion

If all cells are entered during a period before capping at equal probabilities, the distribution of Varroa in cells and on combs should conform to random patterns. In contrast, the numbers of cells containing 0,1,2,... etc. mites was found to differ from poisson distributions. On combs, areas of high and low infestation density are common. This phenomenon is a main source of error in the attempt to estimate the hive infestation from brood samples. The areas differing in infestation rate can be related to areas containing larvae of differing age. Since these comb areas had been invaded by the mites at different times, this apparently reflects considerable fluctuations of the infection rates within short periods.

Among other possible factors, e.g. variable cell attractivity or interactions between Varroa, variable infestation rates may produce the

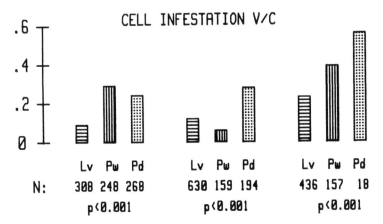

Fig. 3: Infestation rates (Varroa/cell) of capped worker cells (ordinate) containing larvae (Lv), pupae with white eyes (Pw) and pupae with dark eyes (Pd)

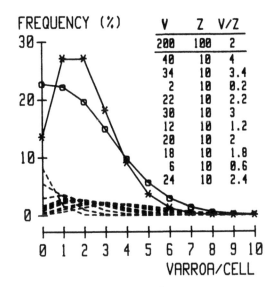

Fig. 4: Model calculations of cell infestation frequencies. *=poisson distribution according to the underlined numbers of Varroa (V) and cells (Z) given in the table. Broken lines: Poisson distributions with random fraktions of Varroa and constant cell numbers according to the table, O= sum curve of the broken lines.

observed deviations of the cell infestation frequencies from a poisson distribution. This can be demonstrated by model calculations as shown in the Fig. 4 example. The total number of mites entering the comb is randomized into 10 unequal portions each infesting 1/10 of the cells. The numbers of cells containing 0,1,2,... etc. mites can then be calculated according to each of the 10 different poissson distributions separately. If these cells are all added up, the resulting distribution shows differences from a poisson distribution resembling the observed differences over a wide range of infestation rates (1).

Interestingly , the reverse procedure, i.e. variable cell availability in combination with constant numbers of infective mites will lead to different results. In particular, a comparable increase of the numbers of low-infested cells will not be produced since this would demand exceedingly high numbers of cells to sufficiently dilute the Varroa portions. The observed variations of the infestation rates of brood comb areas thus apparently do result mainly from substantial short-term fluctuations of the numbers of infektive mites rather than from variations of the numbers of available cells.

REFERENCE

1. FUCHS, S. (1985): Untersuchungen zur quantitativen Abschätzung des Befalls von Bienenvölkern mit Varroa jacobsoni OUDEMANS und zur Verteilung des Parasiten im Bienenvolk. Apidologie 16(4), 343-368

Invasion of honeybee brood cells by *Varroa jacobsoni* in relation to the age of the larvae

S.Fuchs & K.Müller

Institut für Bienenkunde (Polytechnische Gesellschaft), Fachbereich Biologie, J.W.Goethe Universität Frankfurt am Main, Oberursel, FR Germany

SUMMARY

Brood combs from highly infested colonies containing larvae of known age were investigated for Varroa. No mites were found in drone cells or worker cells prior to 170 h since oviposition. Combs were invaded during 60 h and 30 h from capping, respectively. Brood combs may thus be changed among colonies before the 7th day without tranferring Varroa. The attractivity of larvae was investigated in dual-choice tests. Larvae were most attractive shortly after capping, while the attractivity decreased during the pupal stage.

1 Introduction

It is a commonly quoted opinion, that Varroa enters brood cells shortly before capping (1,2), but quantitative data are not available. For practical beekeeping, it is important to know up to which age of the brood combs may be transferred between colonies without carrying mites. In addition, the time course of cell infestation might give some indication of the stimuli involved in the cell entering of Varroa.

2 Results

Dated brood combs, where all eggs had been laid within 8 h in most of the cases, but within 16 h at maximum, were placed within highly infestated bee colonies. The infestation of the cells was then recorded at variable time intervals from oviposition. The relative infestation rate was calculated as percentage of the infestation rate (Varroa/cell, V/C) of capped cells, which was about 1 V/C in worker cells and 5 V/C in drone cells. The resuls from 14 worker combs and 7 drone combs are shown in Fig. 1. In both larva types, no mites were found in the cells if the bee brood was less then 170 h (appr. 7 days) old, and first mites were found in open cells when capping had started in some of the comb cells. With the progress of capping, the infestation of the open cells increased. The period of comb capping took about 30 h in worker combs and about 60 h in drone combs.

In dual-choice tests, two bee larvae were placed at a distance of 3 cm within plastic disposal trays and a Varroa was placed in the middle (35°C, 70% rel. humidity). Young larvae (5-6 d since oviposition) were not attractive in comparison to pupae with white eyes, which were used as reference group in all tests (0:12, p<0.001). Attractivity culminated in the larvae just capped (24:6, p<0.002) and then slowly decreased during the pupal stage (8:24, p<0.01).

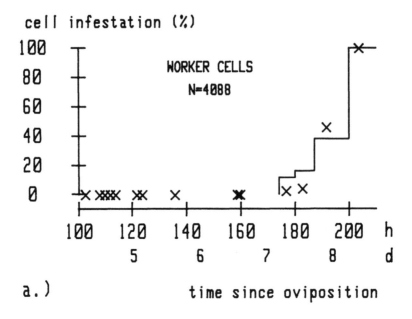

a.)

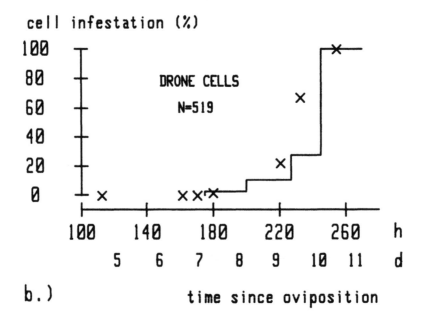

b.)

Fig.1: Infestation of brood combs (x) in % (ordinate) of the infestation of capped cells in worker cells (a) or drone cells (b) in relation to the time since oviposition (abscissa). Solid line: percentage of capped cells

3 Discussion

The results confirm, that cell infestation takes place within a short period before capping of 30 h in worker cells and 60 h in drone cells at most. Worker combs as well as drone combs may thus safely be interchanged between colonies without transferring Varroa mites until the 7th day since oviposition. The low or absent attractivity of young larvae accords with the findings of Rosenkranz et al. (3). The culmination of the attractivity about the time of capping indicates, that the main stimulus for cell entering comes from the bee brood and is linked to its stage of development.

REFERENCES

1. GROBOV, O.F. (1977): Varroatosis in bees. In: Varroatosis a honeybee disease 46-70. Apimondia Publishing House, Bukarest
2. DeJONG, D. (1984): Current knowledge and open questions concerning reproduction in the honey bee mite Varroa jacobsoni. In: W. ENGELS (ed.): Advance in Invertebrate Reproduction 3, 547-552. Elsevier Amsterdam, N.Y., Oxford
3. ROSENKRANZ, P. et al: (1984). Optimal host selektion by reproductive female Varroa jacobsoni. In: W. ENGELS (ed.): Advance in Invertebrate Reproduction 3, 628. Elsevier Amsterdam, N.Y., Oxford

Supported by the Deutsche Forschungsgemeinschaft (Ko 400/4-2)

Natural transfer of *Varroa jacobsoni* among honeybee colonies in autumn

F.Sakofski & N.Koeniger
Institut für Bienenkunde (Polytechnische Gesellschaft), Fachbereich Biologie, J.W.Goethe Universität Frankfurt am Main, Oberursel, FR Germany

Summary

Infested and non-infested honeybee colonies of cordovan ('brown') and Italian bees ('yellow') were placed closely to each other. The exchange of bees among these colonies and the transfer of V. jacobsoni mites was monitored. There was a significant correlation between the number of drifted bees and the number of transferred Varroa mites. Homing foragers of the non-infested colonies also brought substantial numbers of mites.

To observe the drifting of the bees and the transferring of V. jacobsoni, one colony with A.m. ligustica and one colony with A.m.carnica homozygous for cordovan were placed close together. In the surrounding area there were only black bees of some colonies in the vicinity.
In each combination one of the two colonies was highly infested by V. jacobsoni while the other was non-infested and broodless. The non-infested colony contained acaricide impregnated carriers which killed incoming mites at a rate of nearly 90% per 24 hours.
Alltogether 4 pairs of colonies were tested. With these 8 colonies 10 experiments were done at two different locations.
The influx of mites into non-infested colonies was monitored by using two acaricide impregnated boards placed between the frames. Incoming mites were killed by these boards and collected on a plastic sheet on the hive bottom. This was covered with wire gauze to prevent the bees from carrying out the dead mites.

During the observation period (7 hours) all foreign bees at the hive entrance of the non-infested colony were counted. (Fig. 1, A+A'). To determine the infestation of the foraging bees of the infested colony, a sample of bees leaving the hive were caught at the hive entrance (Fig. 1,B).

To estimate the infestation of the foraging bees of the non-infested colony outflying and homing bees were caught separately (Fig. 1,C,D). They were killed in a freezer and the mites were washed out.

Observation of drifting and mite transfer

The expected number of transferred mites is calculated by multiplication of the number of drifted bees into the non-infested colony with the corresponding infestation rate of the outflying bees from the infested colony (Table 1). There is a significant correlation (r= .62; n= 10;p< .05) between transferred mites into the non-infested colonies and the expected number of mites (Fig. 2).

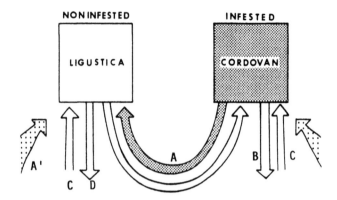

NONINFESTED   INFESTED

LIGUSTICA   CORDOVAN

A'   C D   A   B C

Fig. 1   A: Drifting bees from the neighbouring colony
          A': Drifting bees from other colonies
          B+D: Outflying foragers of own colony
          C: Home coming bees of own colony
               Further explanations see text

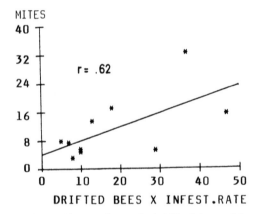

MITES
40
32
24   r = .62
16
8
0
    0   10   20   30   40   50
    DRIFTED BEES X INFEST.RATE

Fig.2 The x-axis shows the number of drifted bees times their
infestation rate, the y-axes shows the number of transferred mites into
non-infested colonies.

In 7 out of 10 experiments the observed number of transferred Varroa
mites into non-infested colonies was mostly higher than the predicted
number of mites (Table 1).
A number of outflying and home coming bees was caught at the entrance of
a non-infested colony to determine the number of mites, carried home by
the own bees. In this experiment a substantial number of homing bees of
the non-infested colony was infested by V.jacobsoni (Table 2).
Chi^2= 14.95, p< 0.01.
There was a correlation between the predicted number of transferred
mites, calculated by the number of drifted bees times their infestation
rate, and the actually transferred number of V.jacobsoni. But the number
of transferred mites was significantly higher than the predicted.
Consequenctly there must be other sources of transfer of Varroa mites
into non-infested colonies.
Homing bees from a non-infested colony brought back substantial numbers
of Varroa mites into their colonies.

82

Table 1   Infestation rate of outflying and home coming bees from a
           'non-infested' colony

TRIAL	INFESTATION RATE OF OUTFLYING BEES VARROA/BEE	NUMBER OF DRIFTED BEES FROM THE INFESTED COLONIES	NUMBER OF COLLECTED V.JACOBSONI	PREDICTED NUMBER OF TRANSFERRED MITES
1	.197	16	8	3.15
2	.197	165	37	32.5
3	.089	62	10	5.52
4	.068	72	10	4.90
5	.173	98	18	16.95
6	.173	43	7	7.44
7	.120	65	5	7.8
8	.083	168	13	13.94
9	.083	185	47	15.36
10	.107	49	29	5.24

Table 2   Bees caught at the hive etrance
             of a 'non-infested' colony

	BEES WITH V.JACOBSONI	BEES WITHOUT V. JACOBSONI
OUTFLYING BEES	1	2344
HOMECOMING BEES	16	2077

Searching for an accurate method to evaluate the degree of Varroa infestation in honeybee colonies

N.Pappas* & A.Thrasyvoulou
Laboratory of Apiculture-Sericulture, School of Agriculture, Aristotle University, Thessaloniki, Greece

Summary

The accurancy of the sampling methods for estimate Varroa infestation of a colony is evaluated in this research. The infestation of a colony from Varroa mites should be given for the bee population and the brood separately. Because of the variable distribution of Varroa mite within the honeybee colony, three samples of 100 adult bees are needed from three different frames of a single body hive. The method of brushing the bees in an alcohol solution (25%), shaking and sifting them through a suitable screen gives precise results. Three samples each of 200 capped cells from three different frames are needed to estimate Varroa infestation of brood. These cells it's recommented to be examined in a crossway direction.

Introduction

Estimation of Varroa infestation of honeybee colonies is necessary to study the behaviour of mite and its population dynamic, to evaluate the effectiveness of different acaricides and the methods of treatments.

Several chemicals were tested for varroacide properties (Marin 1981, Ruttner 1981, Wissen and Maul 1981, Tsellios and Kostarelou-Damianidou 1984 and others), a number of works dealed with population dynamic of the mite (Romaniak and Duk, 1983, Ritter et al, 1984, Ifantidis 1984) but only few works were done to find or evaluate the methods of estimate Varroa infestation (De Jong et al., 1982, Poltev et al. 1981).

The effort of this paper was to etablish a sampling procedure with which the level of Varroa infestation could be accurate evaluated and the results of different acaricides or methods of treatment could be comparable examined.

Material and methods

Estimation of Varroa infestation in bee population. The following two methods were tested.

a) Alive bees were picked up and carefully examined for Varroa mite as described by Tsellios and Kostarellou-Damianidou (1984). After that the bees were killed in 25% alcohol solution and re-examined in the laboratory one by one under stereoscope for missing mites. Mites that fell into solution was also counted. This was repeated in eleven colonies with 200 bees in each trial.

* This paper is part of Pappas's Ph. D. Research project.

Table 1. Estimation of _Varroa_ infestation of adult bees by two
different methods.

Code number of trials*	Picked up alive bees and visual searching		Shaking dead bees in alcohol solution	
	mites found	mites missing	mites found	mites missing
1	28	3	2	0
2	47	9	33	0
3	18	5	23	0
4	21	2	19	1
5	29	2	10	0
6	12	2	5	0
7	14	3	1	0
8	49	3	4	0
9	26	4	2	0
10	20	3	1	0
11	54	3	1	0
Total	318 (89%)	39 (11%)	101 (99%)	1(1%)

* About two hundred bees were used for each trial from different
colonies.

Table 2. _Varroa_ infestation of adult bees collected from
different frames of the same colony.

Number of Frame[+]	Percentage of infestation*		
	colony 1	colony 2	colony 3
1	9.7	7.9	10.5
2	6.8	7.6	20.0
3	5.6	11.7	8.9
4	5.7	9.1	12.6
5	5.5	8.7	13.5
6	8.5	5.8	13.3
7	8.0	9.3	12.8
8	8.4	6.5	19.2
9	5.1	8.9	12.6
$\bar{x}$	7.0	8.4	13.7
s	1.65	1.72	3.65
C.V. %	23.4	20.5	26.6

* 100 bees were collected from each frame
$\bar{x}$:mean, s:Standard deviation, C.V.:coefficient of variation
[+]Number of frames indicates also their position in the colony.

Table 3. Estimation of Varroa infestation of adult bees by
combining the infestation of three frames.

Some possible frames compinations	Varroa infestation		
	Colony 1	Colony 2	Colony 3
1, 2, 3	7.4	9.1	13.1
4, 5, 6	6.6	7.9	13.1
7, 8, 9	7.2	8.2	14.9
C.V.%	5.8	7.3	7.2
1, 4, 7	7.9	8.8	11.9
2, 5, 8	6.9	7.6	17.6
3, 6, 9	6.4	8.8	11.6
C.V.%	11.0	8.1	24.4
1, 2, 9	7.2	8.1	14.4
4, 5, 6	6.6	7.9	13.1
3, 7, 8	7.3	10.1	13.6
C.V.%	5.8	14.0	4.5

* The number of frames and the percentage of infestation
are from table 2.

b) Two hundred bees were brushed into wide-mouth vials containing 25%
alcohol solution, kept for 24 hours in the laboratory and then agitated
mechanically in a shaker for 30 minutes. The mites were sifted and counted
with an appropriate screen as described by De Jong et al.,(1982).

Number of adult bees in each sample indended to estimate the Varroa
infestation level. One hundred bees were collected from each frame of three
colonies (total 900 bees per hive) and their Varroa infestation was found.

Number of capped cells that should be opened. All the sealed cells of
brood nest of colonies with single hive body and 10 frames population,were
opened and examined under stereoscope for Varroa mites. Each infested cell
was marked in a plastic sheet in which the brood area had been previously
traced. For each side of a brood-comb a separate plastic sheet was used
and thus the patterns of Varroa infestation were imprinted. This
facilitates the study of distribution of Varroa mite within the brood nest
of a colony.

To find the distribution of the mites within a brood area of a comb and
the position from which the capped cells should be examined, each plastic
sheet with the marked brood area and the infested cell was divided to zones.
Each zone was formed as the outer edge of six different in sizes rectangles
one inside the other. The inner rectangle was 10x2 cm^2, the second 16x4 cm^2,

Table 4. Variation of Varroa infestation of honey bee
worker brood among different frames of three hives.

Hive	Code number of frame*	number of		% infestation	Statistics
		cells	mites		
A	1r	580	66	11.3	
	1l	290	27	9.3	
	2r	360	101	28.0	$\bar{x}$= 20.4
	2l	565	147	26.0	S= 8.1
	3r	700	163	23.3	C.V%=39.7
	3l	480	126	26.3	
	4r	564	155	27.5	
	4l	170	20	11.8	
B	1r	290	69	23.7	
	1l	360	87	24.2	$\bar{x}$= 16.1
	2r	940	107	11.4	
	2l	1060	152	14.3	S= 5.1
	3r	290	37	12.7	C.V%=31.6
	3l	820	101	12.3	
	4r	900	124	13.8	
	4l	1160	189	16.3	
G	1r	137	25	18.4	
	1l	115	7	6.1	$\bar{x}$= 11.7
	2r	170	27	15.9	S= 6.3
	2l	360	23	6.4	C.V%=53.8
D	1r	1049	71	6.8	
	1l	719	56	7.8	
	2r	701	42	6.0	$\bar{x}$= 7.85
	2l	1095	115	10.5	S= 1.66
	3r	1044	107	10.2	C.V%=21.1
	3l	1764	89	7.6	
	4r	1232	82	6.7	
	4l	1335	112	8.4	
	5r	1719	50	7.0	

* Numbers indicate position of frame in the brood nest (r)
indicates right and (l) left side of the frame.

the third 22x8 cm², the fourth 28x12cm², the fifth 34x16 cm², and the last
40x20 cm².

The center of the brood was focused on the center of these concentric
rectangles which was further treated as brood areas. From the number of
brood cells and the total cells within each brood area the percentage of
infestation was found for each area separately.

After the position of the capped cells had been established, 20,50,
100 and 200 cells were opened from each frame of a colony to find the
number of capped cells that should be in each sample.

Results and discussion

Estimation of Varroa infestation in bee population. Table 1 shows the precision of the two methods that were used to estimate the level of Varroa infestation in eleven colonies. By the picked up alive bees and visually searching method, 39 (11%) of mites out of 357 were missed. On the other hand by shaking dead bees in alcohol solution only one mite was missed. Examine alive bees is time consuming and somehow bias since the observer some times has the inclination to pick up the first infected bee that he sees on the frame.

The subjectiveness, the time that requires and the inaccurancy of the first method make it unsuitable as method of evaluation of Varroa infestation of honeybees. Contrary the second method is very precise, requires a considerably less time and is more reliable.

There are few recommendations on the number of bees that should be included in each sample in order to have a measurement of Varroa infestation of a colony. Poltev et al., (1981) recommented 50-100 bees, Ritter (1981) 200-250 and De Jong and Conçalves (1981) 300-500 bees per colony. Table 2 indicates results of our experiment in which 100 bees were collected from every frame of three colonies. In all colonies there was large variation among Varroa infestation of different frames (coefficient of variation 20.5 to 26.6). There was not relation among the position of the frame and its level of infestation. Frames in the outmost position of the colony may have higher infestation than the middle frames and vice versa. Because of this large variation among frames of the same colony, 100 bees or even more from a single frame could not give an accurate estimation of Varroa infestation of a colony.

On the other hand by pooling the infestation of three frames the coefficient of variation decreases remarkable as it can be seen from Table 3. Almost all the frame combinations gave low coefficient of variation. This lead us to suggest the used of 3 samples, each of 100 bees, for any three frames of a colony as a suitable sample for estimating Varroa infestation in bee population.

Estimation of Varroa infestation in honey bee brood. The variability of infestation among brood of different frames of the same colony, the number of cells that should be examined and their position on the brood area are the principle factors that should be considered to establish an accurate method of determination of Varroa infestation in honey bee brood. Table 4 shows the percentage of infestation of brood of all the frames of four colonies. Code number of frames indicates the position of the frames within the brood nest. As in adult bees, a large variation exist among infestation of different frames of brood of the same colony (coefficient of variation 21.1-53.8%). Differences among frames of the same colony can be as large as 18.7% (Hive A, frames 1 and 2). Differences can be found among the two sides of the same frame (Hive A, frame 4) or among two face to face sides of two frames (Hive A, frames 1 and 2, Hive B frames 1,2). The infestation level obviously is not related to the position of the frame within the brood nest.

Thus, any sample of cells collected and examined from a single frame of a colony, may give the infestation level of that frame and not of the colony. More than one frame is needed for an accurate estimation of the level of infestation of brood. We suggest to examine capped cells from three frames since the coefficient of variation decreases considerably with the most of the combinations.

The so far information of Varroa distribution within the brood area of a frame is rather limited. Ritter (1981) wrote that mites prefer drone

Table 5. Distribution of _Varroa_ mites in worker honeybee brood comb*[1]

Code Numbers*[2]	Number of _Varroa_ mites per 100 cells in each of the brood area that has been examined						Statistics		
Area:	1st	2nd	3th	4th	5th	6th	$\bar{x}$	S	C.V%
A. 1r	13.6	10.3	14.9	8.4	4.4	-	10.3	4.2	40.6
2r	25.3	18.3	27.1	16.6	-	-	21.8	5.2	23.6
21	21.7	19.6	11.6	-	-	-	17.6	11.1	63.0
3r	32.1	31.1	23.8	23.4	14.8	-	25.0	12.4	49.6
31	40.1	24.8	38.4	21.8	11.2	13.6	25.0	13.1	52.4
4r	21.7	24.4	31.5	29.1	29.3	-	27.2	4.0	14.7
B. 11	12.4	18.5	9.2	20.2	22.3	14.6	16.2	5.0	30.7
2r	9.0	9.5	10.8	7.7	10.1	20.0	11.2	4.4	39.7
31	20.4	20.7	20.7	30.1	12.0	8.7	18.8	7.6	40.2
3r	25.6	18.1	18.1	12.2	8.6	10.3	15.5	6.3	41.0
C. 1r	9.4	5.3	3.2	4.0	3.2	-	5.0	2.6	51.6
2r	7.7	4.9	9.1	8.1	4.5	-	6.9	2.1	29.7
D. 11	19.3	14.2	9.0	16.9	13.4	18.4	15.2	3.8	25.0
2r	13.0	15.1	9.8	16.0	21.1	22.2	16.2	4.7	29.0

*1:Brood was divided to 6 areas by the outer edge of six different in sizes
   rectangles one inside the other. The 1st area was the center of the brood.
*2:Capital letters indicate hives,numbers indicate position of frame within
   brood-nest r and l indicate right and left area of the frame.

Table 6. _Varroa_ - infestation % of honeybee brood estimated
by examining a different number of capped cells.

Code number of frame *[3]	Number of cells that were opened				
	all*[1]	20	50	100	200
1r	6.8	25.0	20.0	13.0	6.0
2r	6.0	15.0	14.0	9.0	7.0
3r	10.2	20.0	18.0	15.0	8.5
31	7.6	10.0	14.0	15.0	7.5
41	8.4	4.0	16.0	15.0	13.5
4r	6.7	15.0	14.0	14.0	8.5
5r	7.0	0.0	12.0	13.0	9.0
$\bar{x}$*[2]	7.5a	12.7b	15.4b	13.4b	8.6a

*1. Total number of brood cells are given in Table 4 (Hive D)
*2. Means with different alphabet letter are statistically
    significant different in 0.05 level.
*3. Numbers indicate position of frame within brood nest
    r and l indicate right and left area of the frame.

brood especially round the outer edges of the combs. On the other hand Rosenkranz (1983), when examined the capped brood of 2 skeps found that the distribution of the parasite is rather accidental. Table 5 shows the distribution of Varroa mites within the six brood areas that formed by the six rectangles. In all the frames a wide range of the percentage of infestation existed. Differences beetween the brood areas 1 and 5 of hive A was 28.9%. There are frames that have more mites in the center (brood area 1) than the outer area and vice versa. There are also frames with the most mites in the middle area. These results indicate that the distribution of mites within the brood area is rather accidental and has nothing to do with the center or the periphery of the brood. One way to overcome the problem of this dissimilar distribution of Varroa within the brood nest is to sample sealed cells in crossway manner.

Table 6 shows different groups of capped cells that were opened in a crossway manner, in order to find the number of cells that should be included in each sample indended to use for the estimation of Varroa infestation. Statistically significant differences existed among the infestation of the group of 20, 50 and 100 cells and the total infestation. The differences among the 200 cells-infestation and the total one was not significant. This indicates that a representative sample for an estimation of Varroa infestation of brood, could be the number of 200 cells for each of three frames of a colony.

In conclusion for an accurate estimation of Varroa infestation of bee population 3 samples of 100 adult bees each from any three frames of a single body hive are needed. The method of brushing the bees in an alcohol solution, shaking them and sifting the mites through a suitable screen gives precise results. To estimate the infestation level of brood, again three samples are needed each of 200 sealed cells from three frames. These cells should be opened in crossway manner.

Acknowlegdments

We would like to thank Dr. Ifantidis for his valuable suggestions during the course of this study and also for his comments on the final manuscript.

REFERENCES

1. De Jong D. and L.S. Goncalves 1981. The Varroa problem in Brazil. Am. Bee J. 121 (3): 186-190
2. De Jong D., De Andrea Roma and L.S. Goncalves 1982. A comparative analysis of shaking solutions for the detection of Varroa jacobsoni on adult honeybees. Apidologie 13 (3):297-306.
3. Ifantidis M.D. 1984. Parameters of the population dynamics of the Varroa mite in honeybees. J. Apic. Res. 23 (4) :227-233.
4. Marin M. 1981. Control of Varroa disease with sineacar. Successful results obtained in Romania and other countries. Apiacta 16:53-56.
5. Poltev V., I. Sadov and A.V. Mel'nik 1981. Examination of honeybees for Varroa. Apic. Abstracts 1331/82.
6. Ritter W. 1981. Varroa disease of the honeybee Apis mellifera. Bee World 62 (4): 141-153.
7. Ritter W., E. Leclercq and W. Koch 1984. Observations des populations d'abeilles et de Varroa dans les colonies à differents nineaux d'infestation Apidologie 15 (4): 389-400.

8. Romaniak K. And Duk J., 1983. Seasonal dyanamics of <u>Varroa jacobsoni</u> development in untreated honeybee colonies. Apic. Abstracts 245/85.
9. Rosenkranz P. 1983. Distribution of <u>Varroa</u> females within the honeybee broodnest and consequenses for biological control. Meeting of E.C. experts'group. Thessaloniki 26-28 September.
10. Rutter F. 1981. Field tests of K-79. Apiacta 16:49-52.
11. Santas L.A., M.D. Lasaraki and P.I. Choustoulakis 1984. Comparative testing of various pesticides against Varroatosis of honeybees. Nea melissa 2 (8-9) : 5-12.
12. Tsellios D.E. and M. Kostarelou-Damianidou 1984. The control of <u>Varroa</u> disease by various pesticides Agricultural Research 8: 169-175.
13. Wissen W. and V. Maul 1981. Test procedures of application of formic acid in the cotrol of Varroa disease. Apiacta 16: 66-70.

Session 3
Microbes and laboratory techniques

Chairman: H.Hansen

The incidence of acute paralysis virus in adult honeybee and mite populations

B.V.Ball
Rothamsted Experimental Station, Harpenden, UK

Summary

Studies on the incidence of acute paralysis virus (APV) in European
honey bee (Apis mellifera) colonies infested with the parasitic mite
Varroa jacobsoni have shown that the peak of infection occurs in late
summer. This coincides with a sharp decline of the adult bee and
brood population.
    Mites can acquire APV during feeding on infected hosts and
transmit it to healthy adult bees and brood.
    The possible role of the mite in the initiation and spread of APV
infection of honey bees is discussed.

1. Introduction

The loss of many honey bee colonies throughout the continent of Europe
has been attributed to their infestation with the parasitic mite Varroa
jacobsoni. However its effect on colonies is still poorly understood.
Previous work (2) indicates that death of infested colonies is associated
with acute paralysis virus (APV) infection. This contrasts with findings
in Britain where the mite does not occur and where APV has never been
associated with mortality of field bees.
    The evidence from both field and laboratory studies suggests that
Varroa mites are capable of transmitting APV from severely infected to
healthy individuals, but the factors affecting the initiation of virus
replication and its spread and persistence in both honey bee and mite
populations are unknown. A better understanding of the dynamics of virus
replication in Varroa-infested colonies will contribute to strategies
designed to control the mite and reduce its damaging effect.

2.1 Detection of APV in adult bees

Dead bees from 18 colonies in six different apiary sites in Southern
Germany were collected monthly from May to September and examined for all
pathogens. Previous studies on uninfested colonies in Britain had already
provided much information on the type and prevalence of pathogens causing
natural mortality.
    During this same period Dr. Ritter and his colleagues also estimated
the adult bee, brood and mite populations in these 18 colonies and each
colony was designated Low, Medium or High according to their level of
infestation with Varroa.
    In the colonies with Low mite numbers the adult bee and brood
population increased normally throughout the summer and the viruses present
in the adult bees and their incidence during the year reflected the results
obtained from mite-free colonies in Britain. However in the colonies in
the other two categories of infestation one virus predominated; APV.

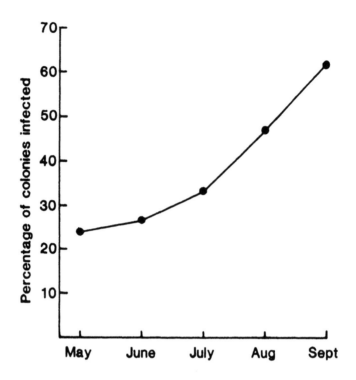

Figure 1. Percentage of German honey bee colonies in which APV was detected in the extracts of dead field bees.

Figure 1 shows the percentage of colonies out of the 18 sampled in which APV was detected by immunodiffusion in the extracts of dead adult bees. There is a progressive increase in virus incidence towards the end of summer so that by September all samples from the colonies in the Medium and High infestation categories contained much APV. This coincided with a sharp decline in both the adult bee and brood population in these colonies reported by Ritter et al., (4) which occurred from July onwards. While the adult bee and brood population declines because of virus infection the mite population remains stable. Consequently the relative infestation of bees and brood by mites increases, virus transmission increases and this leads to rapid collapse of the colony.

Samples of live adult bees were collected from the six German colonies in the High Varroa-infestation category and from four mite-free colonies in Britain at monthly intervals from May to August. A sensitive enzyme-linked immunosorbent assay (ELISA) was used to detect and quantify APV in individuals (1).

Table 1 shows the results from one of the infested German colonies. During the summer there is a progressive increase in both the number of live bees in which APV was detected and in the amount of virus in individuals. Although in May APV was detected in dead bees from this colony the proportion of the adult bee population in which the virus was replicating must have been low as none was detected in the sample of live adult bees. In June five out of eight of the live bees assayed contained detectable amounts of APV and from July onwards virus must have been replicating in most of the adult bee population as APV was detected in all bees sampled.

May	June	July	August
0	0	0.03	0.04
0	0	0.03	0.27
0	0	1.77	2.88
0	0.3	3.27	11.48
0	0.3	5.20	20.42
0	0.14	10.96	24.55
0	22.20	24.55	89.13
0	99.10	79.98	218.80

Table 1. Amount of APV (µg/bee) in individual live honey bees collected monthly from one colony severely infested with V. jacobsoni

APV was not detected by ELISA in honey bees collected from mite-free colonies in Britain. However, infectivity tests confirmed that small amounts of virus were present in the live bees from these colonies, although clearly at levels below the threshold for detection using the ELISA technique.

2.2 Detection of virus in V. jacobsoni

During August sections of comb containing sealed brood were removed from colonies in the High infestation category. Mites were collected from the sealed cells and extracts of the corresponding prepupae and pupae were tested against APV antiserum by immunnodiffusion. Adult female mites were assayed for APV by ELISA.

APV was detected in only five of the 15 mites from cells in which extracts of the corresponding prepupae and pupae failed to react with APV antiserum in immunodiffusion tests. Virus yields were mostly low, near the threshold for detection.

All 15 mites from the APV infected brood (Table 2) contained virus but the amounts varied widely between mites from different cells and between mites in the same cell. The difference in virus content between mites in the same cell may arise in two ways. Firstly, two mature female mites entering the same cell may have acquired different amounts of virus by feeding previously on hosts in which APV was multiplying. Secondly, mite progeny differ in age and would therefore have fed on the APV-infected pupa for varying amounts of time. Progeny ingesting much virus in this way could transmit it to their next host, adult bees, as Batuev (3) demonstrated experimentally.

The ELISA does not distinguish between infective and non-infective virus particles. Moreover, it is unlikely that all of the virus detected in mites would be transmitted to subsequent hosts. However, as some mites contained almost 1µg of virus, equivalent to 10^{10} particles, and 10^2 particles of APV is sufficient to cause mortality by injection into the haemolymph the role of V. jacobsoni as a virus vector must be a significant factor in colony mortality.

3. Discussion

APV normally persists as an inapparent, sublethal infection in adult bees, particularly during the summer months. However, in colonies severely infested with V jacobsoni the virus is a primary cause of both adult bee and brood mortality. The factors affecting the initiation of virus

Brood Cell	Mites	APV (ng/mite)
1	1	219
2	1	708
3	1	178
	2	166
4	1	1259
	2	562
5	1	813
	2	813
	3	275
	4	68
6	1	851
	2	708
	3	468
	4	417
	5	55

Table 2. Amount of virus in individual mites removed from sealed brood
cells. The prepupae and pupae from these cells all gave strong
positive reactions to APV antiserum when tested individually by
immunodiffusion.

replication are at present unknown but it is possible that virus is
released from tissues in which it is normally contained when cells are
damaged by the mite feeding. Virus replication may also be induced
experimentally in adult bees by the injection of foreign proteins and in
nature digestive enzymes secreted by the mite may have a similar effect.
Whatever the mechanism once APV starts to multiply systemically the mite
can then act as a virus vector transmitting APV to other adult bees or
pupae.

The monitoring of virus levels in both honey bee and mite populations
will give further insight into the complex association between V. jacobsoni
and APV and may help to reduce losses by more effective timing of
acaricidal treatments.

REFERENCES

1. ALLEN, M.F., BALL, B.V. WHITE, R.F. and ANTONIW, J.F. (1986). The
   detection of acute paralysis virus in Varroa jacobsoni by the use of a
   simple indirect ELISA. J. apic. Res. 25: 100-105.
2. BALL, B.V. (1985). Acute paralysis virus isolates from honey bee
   colonies infested with Varroa jacobsoni. J. apic. Res. 24: 115-119.
3. BATUEV, Y.M. (1979). New information about virus paralysis.
   Pchelovodstovo 7: 10-11.
4. RITTER, W., LECLERCQ, E. and KOCH, W. (1984). Observations des
   populations d'abeilles et de Varroa dans des colonies a different
   niveaux d'infestation. Apidologie 15: 389-400.

Transmission of honeybee viruses by *Varroa jacobsoni* Oud.

F.P.Wiegers

Department of Entomology, Wageningen Agricultural University, Wageningen, Netherlands

Summary

Varroa jacobsoni Oud. can transmit acute paralysis virus (APV) from one honeybee pupa to another with an efficiency of about 70 %. There is no latent period in the virus transmission. Virus transmission can occur from within 25 minutes up to 48 hours after aqcuisition. Most mites only transmit virus once after an acquisition period. Probably there is no virus multiplication in the mite. Possible transmission mechanisms are discussed.

1. Introduction

BATUEV was the first to demonstrate an association between acute paralysis virus (APV) and the parasitic mite Varroa jacobsoni Oud. in Russia (6). In Germany the same association was found by BALL (5).

BATUEV demonstrated furthermore that the mite could transmit APV from one adult bee to another in the laboratory (6). ALLEN et al. found APV in adult female mites in naturally infested honeybee colonies and thus made it plausible that Varroa can also act as a vector of APV in nature (1).

In the present study an attempt was made to elucidate the mechanism by which virus-transmission occurs, by assessing the influences of the length of the aqcuisition and infection access periods and the addition of a starvation period.

2. Material and Methods

Virus and antiserum preparation

APV was purified from inoculated pupae as described by BAILEY and WOODS (3). The original inoculum was donated by Dr.B.V.Ball from Rothamsted Experimental Station, Harpenden, UK. Antiserum was prepared in a rabbit by giving two intramuscular injections of 1 mg of purified virus in 1 ml of 0.01M phosphate buffer pH7.0 emulsified with an equal volume of Freund's complete adjuvant.

Mites

Adult female mites were collected from worker brood cells. In mites collected from the same colonies ALLEN et al. could not demonstrate any APV (1).

Pupae

Honeybee pupae were collected from colonies having no or a low Varroa infestation. Only completely white pupae (up to 2 days after the pupal molt (9)) were used.

Transmission experiments

The transmission experiments were carried out in glass tubes with an inner diameter of 6mm and a lenght of 25mm. The tubes were closed with a cotton wool plug (Fig. 1).

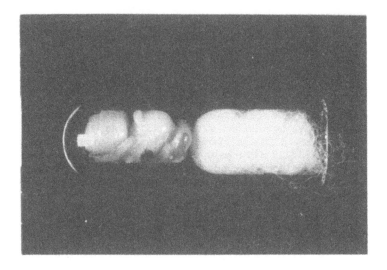

Fig. 1: Glass tube, with a honeybee pupa and a <u>Varroa</u> mite, as used in the transmission experiments.

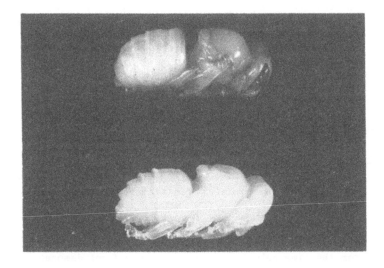

Fig. 2: A healthy (above) and an APV-infected (below) honeybee pupa, 5 days after the pupal molt. The infected pupa was injected with APV, the first day after the pupal molt.

The tubes were put together on a shallow tray placed over a solution of 12% glycerol in water with 0.02M sodiumazide, to maintain a high humidity, enclosed in a larger tray incubated at 30°C.

Starvation of the mites was effectuated by keeping 10 mites in a glass tube for 24 hours.

Weighing of adult female Varroa mites was performed on a Cahn automatic electro balance, model 4700. Pupae injected with 10^{-6} ug of APV were incubated for 2 days at 30°C. Mites were kept during different acquisition access periods (AAP) on these pupae. After the AAP the pupae were frozen.

The pupae on which the mites stayed for the infection access period (IAP) were incubated at 30°C for 5-10 days and then frozen. APV infected pupae could be distinguished from healthy ones because of lack of pigmentation (Fig. 2). In addition, all the pupae were checked for virus multiplication by means of a direct enzyme linked immuno-sorbent assay (ELISA) according to Clark and Adams (6). For this ELISA pupae were extracted in 900 ul of phosphate buffered saline, containing 0.05% Tween-20 (PBS-T). This extract was diluted ten times with PBS-T. Since there was a clear difference between the reactions of extracts of healthy and infected pupae, the reactions could be assessed visually. There were no intermediate reactions.

3. Results

The results of the transmission experiments are summarized in Tables I and II. Table I gives the cumulative efficiency of APV transmission by the mites after a repeated stay on healthy pupae. Table II gives the frequency distribution of APV transmissions.

Table I demonstrates that Varroa mites can transmit APV from one pupa to another with a cumulative efficiency of 69.4 - 89.5 % after 4 IAP. The highest percentages of infected pupae are always found after the first IAP. For non-starved mites, this percentage increases with the length of the acquisition access period. From the second IAP on, the increase of the percentage of infected pupae steadily declines. However after the 3rd or 4th IAP the increase is in some cases higher than after the previous period. Because an increasing percentage of infected pupae is still found after 36 hours, it appears that some mites transmit APV for the first time after 36 hours. No APV-transmission was found after 48 hours.

Of the mites that transmit virus the majority does this only once, with an efficiency which increases with the length of the AAP (Table II).

Table I: Cumulative efficiency (%) of APV-transmission by adult female Varroa-mites after different acquisition (AAP) and infection (IAP) access periods.

experiment	n	condition of the mites	lenght of AAP	length of IAP	number of IAP				
					1	2	3	4	5
1	36	starved 1 day	0.75h	6h	50.0	58.3	61.1	69.4	72.2
2	20	not starved	1h	12h	30.0	45.0	70.0	75.0	-
3	19	not starved	12h	12h	47.4	68.4	84.2	89.5	-
4	16	not starved	24h	12h	62.5	68.8	68.8	75.0	-

Table II: Transmission frequency distribution of adult female Varroa-mites at different acquisition (AAP) and infection (IAP) access periods.

experiment	n	condition of the mites	length of AAP	length of IAP	transmission frequency distribution					
					0	1	2	3	4	5
1	36	starved 1 day	0.75h	6h	27.8%	36.1%	19.4%	13.9%	2.8%	0.0%
2	20	not starved	1h	12h	25.0%	45.0%	30.0%	0.0%	0.0%	-
3	19	not starved	12h	12h	10.6%	57.9%	21.9%	5.3%	5.3%	-
4	16	not starved	24h	12h	25.0%	56.2%	18.8%	0.0%	0.0%	-

Starvation of the mites could only be carried out in groups. Most of the mites died when starved individually. Although the IAP of the starved mites was only 6 hours the percentage of infected pupae in the first IAP is relatively high when compared to non-starved mites

Little is known about the feeding behaviour of the mites. Because we were unable to observe food uptake of the mites, we tried to establish this by weighing the mites, as a non-destructive method of measuring food-uptake After a starvation period of 24 hours, the weigth of the mites dropped from 400-500 ug to 250-350 ug. After placing the mites on pupae, an increase in weight of up to 50 ug in 25 minutes was found. However, no correlation could be demonstrated correlation between weigth increase and virus-transmission. Consequently the measured weigth increase is not caused by food uptake, or virus-transmission is not related to the amount of food consumed. APV-transmission was even found in an IAP as short as 25 minutes after an AAP of 25 minutes.

4. Discussion

Direct ELISA is a sensitive serological method which does not consume much material. Although the preparation for ELISA takes more time than a gel-diffusion test, it is less labour-consuming, once adapted to the desired antibody-antigen system. Although the more complicated and cost-consuming indirect ELISA is more sensitive it is not a necessity for this purpose (2). As Allen et al. already stated an indirect ELISA would be usefull when more different viruses are studied, or when a larger sensitivity is required, e.g. in case of the study of APV in mites (1).

Because of the fact that after 36 hours, some mites still can transmit APV for the first time, it seems unlikely that transmission of APV is merely effectuated by contamination of the mouthparts. In addition few mites infected a pupa with APV in more than one IAP. If the virus was transmitted by contamination of the stylets, the virus would probably be removed during feeding and subsequently not be available for a second infection. On the other hand, we found no indication for the possibility that virus is circulated in the mite. A circulation of virus would result in a latent period in which no virus transmissin occurs. This circulation can probably not be effectuated within 50 minutes, being the sum of the shortest AAP and IAP that were found. Therefore another mechanism is more likely. It is suggested that the virus is ingested with the hemolymph of the bee and is stored in the alimentary duct. Infection with that virus could take place by regurgitation of gut content just before feeding, as is found in beetles (7). Detailed studies of feeding behaviour of the mite therefore seem necessary.

SADOV measured a weigth increase of 80-140 ug per 2 hours, which is in agreement with the 50 ug increase per 25 minutes reported here (11). TEWARSON however calculated a consumption of 1.5 ul per 48 hours, which is much lower. PETROVA et al. calculated from oxygen consumption a quantity of food uptake which is also less than the weigth increase reported here (10). What PETROVA et al. in fact calculated was nutrient uptake in the hemolymph of the mites, whereas the results of SADOV and the results reported here represent hemolymph uptake. The 1.5 ul that TEWARSON reported can contain up to 10^{10} virus particles. With this quantity of virus a large amount of infections can be caused. This implies that the greater part of the ingested virus is not available for transmission.

Since no transmission was found 2-5 days after the aquisition period, and most mites only transmit APV in one IAP, multiplication of the virus in the mite is not likely, at least not to such a degree that more virus will be available for transmission.

The increase of transmission efficiency caused by a prolongued AAP can be explained by an increased chance of the mite feeding on the infected pupae and thus an increased change of acquiring APV. The addition of a starvation period urges the mites to feed more on infected pupae, which also results in an increased change of acquiring virus.

Black queencell virus (BQCV), another honeybee picornavirus, could not be transmitted by Varroa jacobsoni in previous experiments. At this moment Varroa is tested by the author for its ability to transmit chronic paralysis virus (CPV), a honeybee virus that does not belong to the picornavirus group and has been the cause of mortality in honeybee colonies (4).

Assuming that the mechanism of virus transmission in adult bees is the same as in pupae, the results found in the transmission experiments imply that in nature mites, after having acquired virus from an infected adult bee, could transmit it to other adults bees or brood, as was already found by BATUEV and ALLEN et al.(6,1). In that way APV can become one of the main causes of death of honey bee colonies after an infestation with Varróa jacobsoni, as was found by Ball (5).

5. Acknowledgements

The author wishes to thank drs. J. Beetsma for fruitfull discussions and critically reading the manuscript. Furthermore the author wishes to express his gratitude to the E.C. for financially supporting this work (contract nr.1912) and to the Board and Mr. H.J. Lutke Holzik of the Technical University Twente for all facilities placed at our disposal to carry out our studies with Varroa infested colonies at the campus in Enschede.

6. References

1. ALLEN, M.F., BALL, B.V., WHITE, R.F. and ANTONIW, J.F. (1986). The detection of acute paralysis virus in Varroa jacobsoni by the use of a simple indirect ELISA. J.apic.Res. 25, 100-105.
2. ANDERSON, D.L. (1984). A comparison of serological techniques for detecting and identifying honeybee viruses. J.Invertebr.Pathol. 44, 233-243.
3. BAILEY, L. and WOODS, R.D. (1974). Three previously undescribed viruses from the honeybee. J.gen.Virol. 25, 175-186.
4. BAILEY, L., BALL, B.V., PERRY, J.N. (1981). The prevalence of viruses of honeybees in Britain. Ann.appl.Biol. 97, 109-118.

5. BALL, B.V. (1985). Acute paralysis virus isolates from honeybee colonies infested with Varroa jacobsoni. J.apic.Res. 24, 115- 119.
6. BATUEV, Y.M. (1979). New information about virus paralysis. Pchelovodstvo 7, 10-11.
7. CLARK, M.F. and ADAMS, A.N. (1977). Characteristics of the microplate method of enzyme-linked immunosorbent assay for the detection of plant viruses. J.gen.Virol. 34, 475-483.
8. FULTON, J.P., SCOTT, H.A., GAMEZ, R. (1980). Beetles. In: Harris, K.F. and Maramorosch, K. Eds. Vectors of plant pathogens, Academic Press, New York. 115-132.
9. JAY, S.C. (1962). Colour changes in honeybee pupae. Bee World 43, 119-122.
10. PETROVA, A.D., BYZOVA, Yu.B., TATSII, V.M., EMEL'YANOVA, O.Yu. (1982). Metabolic expenditures of Varroa jacobsoni Oudemans, 1904 (Mesostigmata, Varroidae) - an ectoparasite of the honeybee. Doklady Biological Sciences 262, 115-118.
11. SADOV, A.V. (1976). Izuchenie samki varroa. Pchelovodstvo 8, 15-16.
12. TEWARSON, N.C. (1983). Nutrition and reproduction in the ectoparasitic tic honeybee (Apis sp.) mite Varroa jacobsoni. Dissertation, Eberhard-Karls-Universität Tübingen.

The antibacterial response of haemolymph from adult honeybees (*Apis mellifera* L.) in relation to secondary infections

P.R.Casteels, D.Van Steenkiste & F.J.Jacobs
Laboratory of Zoophysiology, State University of Ghent, Ghent, Belgium

Summary

Survival tests revealed that adult workerbees possess an efficient defense system, which enables them to survive injections with large numbers of non-pathogenic bacteria.

Although there exists an apparent similarity between the defense system of honeybees and that of other insects, the cellular response in honeybees is not sufficient to eliminate E.coli-bacteria from the haemocoel. Due to this deficiency, the bacteria were able to multiply rapidly during the first hours post-inoculation (P.I.). Between 5 and 10 hours P.I., a humoral response reduced drastically the number of bacteria.

Gel electroforesis (native conditions), linked with bio-assays, demonstrated an enhancement of the lysozyme titer and the induction of a cecropin- and attacinlike activity.

1.Introduction

The mite, Varroa jacobsoni, has been held directly responsible for the death of a considerable number of bee colonies although secondary infections may have contributed to the pathogenic effects of Varroa on the honeybee.(1)
Initial studies on Varroa-disease dealt mainly with the parasite itself.

In order to understand the influence of secondary infections however, a complete knowledge of the host's defence-mechanisms is necessary.

The studies presented here were undertaken to investigate the naturally occurring resistance of the adult honeybee to bacterial infection of the haemocoel.

2.Materials and methods

2.1.Organisms : * Honeybees, obtained from healthy
        colonies (A. mellifera carnica F1) were kept
        in Liebefeld cages and fed on sugarsolution (1/1).
        Young bees received pollen during the first 10 days.
        Injections were carried out with a 5 μl syringe in
        sterile conditions.
        The haemolymph was withdrawn from the abdomen.

        * Bacteria. The standard assay organism
was Escherichia coli NCTC 9001. This and the
following organisms were kindly supplied by Dr. M.
GILLIS, State University Gent, Lab Microbiol. and
Microbiol. Genetic.) :
    Erwinia carotovora ssp carotovora NCP 312
    Serratia marcescens ATCC 13880
    Micrococcus lysodeikticus LMG 4050
In addition E.coli K 514 provided with an ampicil-
lin resistance factor (pUC 18) was employed (supplied by M. ZABEAU, PGS). Unless otherwise stated
E.coli NCTC 9001 was used. These strains were grown
aerobically at 33°C on LPGA (2)
Bacterial concentrations were determined in a
counting chamber and living suspensions were veri-
fied by plate count.
Bacterial suspensions were made with PBS (phosfate
buffered saline : 0,8 % NaCl, 0,02 % KH$_2$PO$_4$, 0,02 %
KCl, 0,115 % Na$_2$HPO$_4$) 0,15 M pH=7,2.

2.2. Survival tests : data were analysed, using the SPSS,
     according tot NIE and HULL (3).

3. Results

3.1. Investigations on the naturally occurring resistance
     against bacterial infection of the haemocoel.

    Plating of lymph, obtained from 100 individual bees, on
rich medium yielded no bacterial growth, strongly suggesting that the haemolymph of healthy adult honeybees is
sterile.
    In a series of experiments we injected a predetermined
number of organisms of different bacterial strains.
The honeybeemortality was noted daily. The results for two
bacterial strains are given in fig.1.

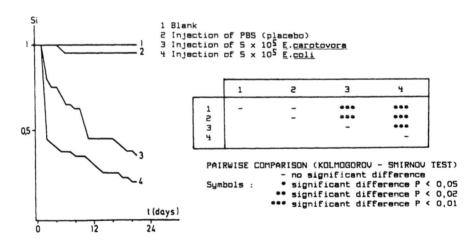

Fig.1 : Survival in days (groups of 50 bees)

S.marcescens (not shown) proved to be fully pathogenic :
inoculations with less than 10 bacteria were lethal after
24 h E.coli and E.carotovora were non pathogenic but the
infections resulted in a lower life expectancy.

3.2.Immunisation of adult workerbees against E.coli.

Injection of foreign particles into individuals of
several insect species elicits an aquired humoral immunity
to subsequent bacterial infection (4).
A similar response was obtained in the honeybee
following survival tests.
Bees were treated by wounding (imitating the Varroa mite),
injection of PBS or by inoculation with an immunising dose
of E.coli. After 48 h a high dose (5×10^5 cells) of E.coli
was injected ; the control group received only this second
injection.
The results shown in Fig.2 show a significant difference in
the survival rate between the treated bees and the control
group. The placebo and the group which had been wounded in
the abdomen were probably activated by introduction of
particles, such as hairs and pollen, as suggested by DUNN
(4).
The series 2 and 3, which received a prior injection of an
immunising dose of E.coli, aquired a moderate protection
resulting from the first injection.
This first dose not only activated the humoral response
but, as already mentioned, also reduced the life expectancy.
In an analogous experiment where the second injection
took place 5 h after the first treatment, the groups
displayed a comparable mortality rate (data not shown).

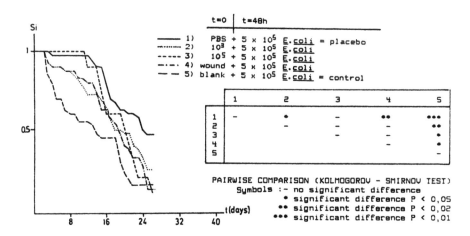

Fig.2 : Survival in days (groups of 30 bees)
S_i = cumulative survival rate

These experiments revealed that adult honeybees are resistant to certain bacterial infections and that they can aquire an immune state.
The role of cellular and humoral factors is further elaborated in the following sections.

3.3. Elimination of foreign material from the haemocoel.

Insect haemocytes (granulocytes GR and plasmatocytes PL) are extremely efficient in removing foreign particles (such as bacteria) from the haemocoel by phagocytosis or nodule formation (5).
After injection of 10^5 E.coli cells a rapid increase in the bacterial cell numbers was followed by a drastic reduction (after 5-10 h) to a residual level of infection. (Fig.3)

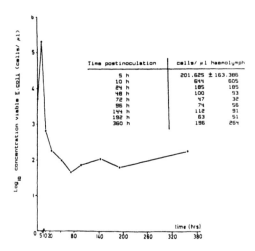

Time postinoculation	cells/ µl haemolymph	
5 h	201.625	± 163.386
10 h	644	605
24 h	185	185
48 h	100	93
72 h	47	32
96 h	74	56
144 h	112	91
192 h	63	51
360 h	196	264

Fig.3 : Concentration of viable bacteria in haemolymph at times following injection.

Moreover honeybee-haemocytes were unable to eliminate completely inert substances such as chinese ink (CHI), Sephadex tracer (∅ 15 µm), latex beads (∅ 0,9 µm) and active coal from the haemocoel.
Although considerable amounts of CHI were phagocytised by the GR, significant amount of ink remained acellularly and was concentrated in the region of the aorta-coils (between thorax and abdomen) and loosely accumulated against pericardial cells.
The defective cellular response was also demonstrated in vitro : whereas GR were able to phagocytise heat-killed E.coli, we were unable to demonstrate phagocytosis of viable bacteria (Fig.4).

108

Fig.4 : In vitro culture of granulocytes inoculated with :
A : Living E.coli ; B : heat-killed E.coli

3.4.Lymph as medium for bacterial growth

As haemocytes were found not to be of prime importance
in elimination of viable bacteria from the haemocoel,
another process had to be responsible for the reduction of
cell number and inhibition of bacterial growth in the
haemolymph.

Haemolymph from 75 adults bees : blank or infected (48
h postinoculation with 3×10^4 E.coli) was pooled.
After centrifugation the lymph was added to PBS to a final
concentration of 10 %, sterilised by filtration (0,22 μm
filter) and assayed for antibacterial activity by inoculation with 10^3 bacteria. Bacterial development in this
incubation medium was monitored by measuring the optical
densities and by plate counts.
The result given in Fig.5 show a clear difference in the
growth curves of the bacterial in tho presence of immune
and blank lymph.

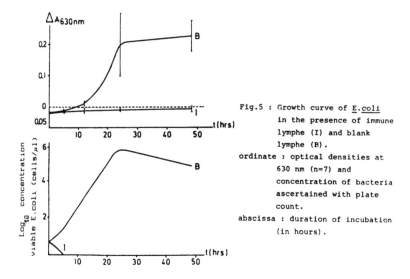

Fig.5 : Growth curve of E.coli
    in the presence of immune
    lymphe (I) and blank
    lymphe (B).
ordinate : optical densities at
    630 nm (n=7) and
    concentration of bacteria
    ascertained with plate
    count.
abscissa : duration of incubation
    (in hours).

The propagation of the E.coli bacteria is thus
inhibited by a humoral factor in the immune lymph.

3.5.Assay for antibacterial activity

MOHRIG and MESSNER (6) determined the lysozyme titer of blank honeybee-lymph with M.lysodeikticus as lysozyme indicator.

We demonstrated by mean of an analogous agar diffusion assay that the lysozyme titer increases significantly in response to wounding or injection by foreign material (results not shown).

This immune lymph was also bactericidal against E.coli; S.marcesens which was fully pathogenic was not susceptible.

As lysozyme is specifically active against Gram+ bacteria, other humoral factors must also be involved since the Gram- E.coli was also inhibited.

These factor(s) were heat-resistant (30 min 100 $^{\circ}$C) and appeared in haemolymph between 5 and 10 h P.I.

Preliminary results showed an analogous response with larvae (worker and drone brood).

3.6.PA-gelelectrophoresis of haemolymph

Humoral factors were further studied by mean of gelelectrophoresis.

We compaired haemolymph samples of blank and immune honeybees by Isoelectric Focusing (IEF) (pH= 3,5 - 10). After seeding the gel with E.coli, bactericidal activity was observed in the basic region (pH>9).

In a next step we used a Cathodic Discontinuous PAGE-system (pH=4, PA=15 %) to separate the basic proteins. On overlaying the gel with bacteria, three types of humoral bactericidal factors were observed : lysozyme and two additional factors directed against Gram- bacteria.

According to DUNN (3) these two factors represent cecropin- and attacinlike activity (Fig.6)

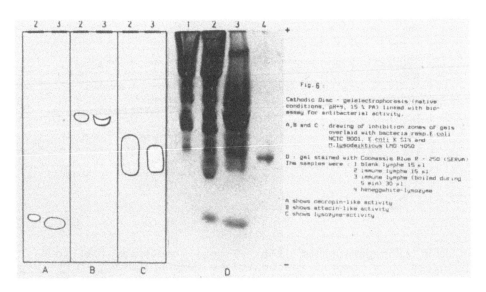

Fig.6 :

Cathodic Disc - gelelectrophoresis (native conditions, pH=4, 15 % PA) linked with bio-essay for antibacterial activity.

A,B and C : drawing of inhibition zones of gels overlaid with bacteria resp.E.coli NCTC 9001, E.coli K 12% and M.lysodeikticus LMG 4050.

D : gel stained with Coomassie Blue R - 250 (SERVA) The samples were : 1 blank lymphe 15 µl
2 immune lymphe 15 µl
3 immune lymphe (boiled during 5 min) 30 µl
4 heneggwhite-lysozyme

A shows cecropin-like activity
B shows attacin-like activity
C shows lysozyme-activity

4.Conclusion

In insects, the exoskeleton forms a primary defence against pathogens. Organisms which pass this barrier will then encounter the cellular and humoral responses in the haemolymph. The exoparasitic mite Varroa jacobsoni is considered to transmit microbial organisms during feeding.

We have investigated the influence of bacterial infections of the haemocoel on adult honeybees.

We concluded that the cellular response of the honeybee is unable to eliminate (non) pathogenic bacteria, resulting in a propagation of the invading organisms.

This resulted in a lower life expectancy of the infected honeybee. Persistance of the residual number of viable bacteria (pathogenic & non-path.) may be a significant factor in the rapid spread of an infection by the mite.

The humoral response was shown to be responsible for the drastic reduction in the amount of viable, non pathogenic bacteria and proved to be analogous to that of other insects : increased lysozyme titer and an induction of cecropin- and attacinlike activity.

References

1.GRIFFITHS D.A., GRAY J. and PEGAZZANO F. (1983) in :
   CAVALLORO R. pp 79-83, A.A. Balkema, Rotterdam
2.VANTOMME R., SWINGS J., GOOR M., KERSTERS K. and DE LEY J.
   (1982) Phytopath. Z., 103 : 349-360
3.NIE H.N.and HULL C.H. (1975 and 1981) Mc Graw-Hill
   Book Compagny, New York
4.DUNN P.E. (1986) Ann. Rev. Entomol., 31 : 321-389
5.SALT G. (1970) Cambridge University Press, London
6.MOHRIG W. und MESSNER B. (1968) Biol. Z. bl.,
   87 : 439-470

Acknowledgement
This work was supported by a grant from the Belgian Institute for Encouragement of Scientific Research in Industry and Agriculture (IWONL) to one of us (P.R.C.).

111

Some preliminary observations on the behaviour of *Varroa jacobsoni* Oud. on its natural host under laboratory conditions

F.Chiesa & N.Milani
Istituto di Difesa delle Piante, Università degli Studi, Udine, Italy

Summary

The results are given of some preliminary tests on the breeding of **Varroa jacobsoni** in the laboratory on its natural host. A highly successful technique is described for the rapid transfer of bee larvae from the comb to an artificial environment. The mites survived well in the artificial environment, but great difficulties were encountered in regard to reproduction. The phenomenon of aggregation shown by the female adult of **Varroa jacobsoni** is illustrated and its biological significance is discussed.

1. Introduction

The particular habitat of **Varroa jacobsoni** makes the study of its biology difficult. Actually, the large number of mites needed for different biological tests would presuppose the availability of strongly infested beehives, what makes their survival precarious.

In the countries most affected by the spread of this mite, research has been carried out on breeding it on its host outside the natural environment; the results are little known as they are often published in languages understood by few and in journals with a limited distribution in Western Europe.

Sakai, Tacheuchi and Hara (12) and Akimov and Piletskaia (1) discovered the thermohygrometric conditions most favourable for the reproduction of **Varroa jacobsoni**; Avdeeva (2,3) found a way of triggering egg laying in mature females. Interesting contributions have also been made by Cavicho Issa (5), and Cavicho Issa and Gonçalves (6) who noted the large difference in the rate of reproduction of mites bred **in vitro** on drone and on worker larvae.

In the studies mentioned above breeding was done in glass test tubes or similar containers, so only a single larva was available to one or more mites. These tests were only partially successful and showed the dependence of the mite on the bee as a source of food and also the strict dependence on the hive environment.

Thus alongside the research in progress at the Istituto di Difesa delle Piante of Udine University, to find an artificial diet for rearing **Varroa jacobsoni**, we also studied the ability of the parasite to adapt to an artificial environment in which its nutritional needs were satisfied by the presence of the host. To this end comparisons were made between artificial environments more or less similar to the natural one (plastic or wax plates, pieces of comb, etc.). Numerous preliminary tests were carried out in part to detect technical difficulties which might be met with while using these apparatus. A summary of the results is given here.

2. Techniques for transferring bee larvae to artificial environments.

Bee larvae which had almost reached the sealing stage (as they do not need further feeding) were transferred to the artificial environments.

Combs containing larvae at this stage of development were taken from Dadant-Blatt type hives during the months of May to September 1986. The combs were kept at a temperature of 34±1°C and a relative humidity of 85% in a vertical, or almost so, position in the dark. The larvae which have almost reached sealing emerge from the cells within 2-4 hours and can be collected on pieces of soft paper placed beneath the combs, or alternatively they can be gathered just before they emerge from the cells, but in such a way as not to subject them to any pulling.

The larvae were then transferred to an artificial environment consisting of polystyrene plates used for the micro-ELISA (enzyme-linked immunosorbent assay) test (Dynatech) which contain a series of 96 almost cylindrical wells, 6.5 mm in diameter and 11 mm deep, resembling worker bee cells. The micro-ELISA plates were chosen for their transparency, as the behaviour of the mites in the presence of a relatively high number of hosts could be observed and the conditions were closer to natural ones in respect to test tube breeding (2,3,5,11), even if the risk of contamination by moulds was much higher.

In the preliminary test on the applicability of this method 96 worker larvae taken from the same comb were transferred to a micro-ELISA plate. They were left to complete their development at 34±1°C and 85% RH (*) in the dark and observed on alternate days over a 10 day period. Seventy-six larvae were transferred using the above described technique; of these 69, that is 91%, reached the adult stage in the same length of time as is needed to develop under natural conditions and their morphology was completely normal. On the other hand, of the remaining 20 larvae, extracted from their cells by the normal technique, with every possible care, only 5 (25%) reached the adult stage. The mortality observed in both groups, particularly at the larval stage, was mainly due to the development of moulds.

(*) The RH is higher than inside the hive, where, during the good season, it does not rise above 60% (4); the conditions are those shown to be the most favourable for the breeding of **Varroa jacobsoni** (1,12), and presumably are near enough to those existing inside the operculate cells where the mites reproduce.

During the following transfers in which the larvae were artificially infested by adult female Varroa, the percentage of bees reaching the adult stage was somewhat variable, because the presence of mites favours the spread of fungal infections (Fig. 1-6).

The above described technique was the only practicable one for a rapid transfer of larvae with a high percentage of success (the transfer of 100 larvae takes about 5 minutes) whereas the manual extraction of larvae during later stages of development often causes damage to the delicate cuticle of the insect and in addition requires a lot of attention and time.

3. Laboratory rearing of **Varroa jacobsoni**.

Varroa jacobsoni was bred outside the hive on its natural host, in incubators at 34±1°C and 85% RH in the dark; bee larvae and the mites were housed in different types of cells. The following were tried: a) the micro-ELISA plates described above; b) by melting virgin wax plates were prepared 20 mm in thickness containing holes 6.5 mm in diameter and 11 mm deep for holding worker bee larvae and 8 mm in diameter and 13 mm deep for drone larvae; c) pieces of bee comb.

The bee larvae were transferred to the micro-ELISA plates and to those made of wax using the technique described above; the mites have been taken from broods of infested hives.

3.1. Behaviour of **Varroa jacobsoni** in the presence of its natural host.

Bee larvae transferred to micro-ELISA plates and to wax plates have been infested with mites and maintained in an incubator until the adult bees hatched out. Two portions of naturally infested comb containing drone brood, from which the operculum had been removed, were kept under similar conditions. A plate smeared with vaseline was placed over each plate to trap escaping mites; the remaining mites were then counted. The results are shown in Table I.

Confinement due to the presence of an operculum is not the only factor keeping the mites on the host; following unsealing almost all the

Table I - Number of mites remaining on the bee larva or pupa in different environments.

Environment	No. of bee larvae	Initial No. of mites	1	2	3	4	5	6	7	8
				No. of mites remaining on day						
ELISA plate 1	96	37				18		5		3
ELISA plate 2	96	47	26			17			9	7
ELISA plate 3	96	35	23	23	23	23			22	21
WAX plate	60	20	15	14	14	14			13	13
Comb with larvae		36	35	35	35	35			35	35
Comb with pupae		87	84	84	82	82			81	81

parasites remained in the two portions of naturally infested comb, while mites kept in artificial environments, especially during the first few days showed a strong inclination to leave the host. It is not easy to explain the results from the different micro-ELISA plates which were obtained during different periods of the year (the beginning of summer for plates 1 and 2, the beginning of autumn for plate 3), so an hypothesis on the different behaviour of the mite at various moments of their annual cycle, must not be discarded.

3.2. Aggregation of **Varroa jacobsoni** on bee larvae.

During the previous tests the tendency of adult females of **Varroa jacobsoni** to gather on certain bee larvae was noted; a similar behaviour occurs also under natural conditions.

To understand this phenomenon two micro-ELISA plates were prepared with 96 worker larvae, and artificially infested with 100 and 50 mites respectively. The plates were covered with a lid (stainless steel net, 100 μm mesh). The distribution of the mites on the larvae was checked every day for the next 4 days.

The distribution was strongly aggregated in the plate with a higher density of mites (Fig. 7-10), deviating significantly from a Poisson distribution (P <.025) and in a closer agreement with a negative binomial distribution, with **k** varying between 0.72 and 1.03 on the different days. The aggregation was less evident in the plate with a lower density, from the third day only a highly significant deviation in regard to the Poisson distribution was noted.

In addition it was observed that mites tend to infest contiguous larvae. The possible pairs of adjacent larvae were divided into three classes: **a**) both larvae infested; **b**) one larva infested and the other free; **c**) both uninfested. The distribution of the larval pairs within these classes is significantly different from what one would expect on the basis of the percentage of infested larvae, if infestation occurred by chance. In fact in all cases, except one, there was an excess of classes **a** and **c** ($\chi^2_2 > 6$, and P <.005). It was also observed that on successive days the aggregation could occur on different larvae and that, starting on the second day, an edge effect becomes evident with mites gathering on peripheral larvae.

This aggregation of mite females favours exogamy since copulation between mites appears to occur within the bee cells, where the male remains confined (7). Further research is necessary to determine this phenomenon quantitively in nature.

3.3. Survival of **Varroa jacobsoni** in laboratory conditions.

Bee larvae transferred to an artificial environment were artificially infested with mites taken from a sealed brood. By means of a 100 um mesh nylon net the mites were confined individually in the cells each containing a bee larva. Both the number of bees which reached the adult stage and the number of surviving mites were noted. As a control, portions of artificially infested (through a small hole in the operculum, which was

116

immediately closed) and naturally infested sealed comb were kept in the same incubator.

The results are summarized in Table II.

The mites survive well; death is connected to that of the host which in turn is mainly due to moulds. Results obtained using micro-ELISA or wax plates were almost equivalent; the failure in regard to drone larvae is due to the greater difficulty of transferring them from their natural environment; this was observed also by Avdeeva (3).

During successive transfers of mites onto worker larvae in wax plates the survival time increased reached beyond a month for almost a third of them (Table III).

However, in spite of the relatively good results obtained in regard to even prolonged survival, eggs were laid only sporadically in all the artificial environments examined.

3.4. Egg laying of **Varroa jacobsoni** in laboratory conditions.

As can be seen from the literature, even in nature some adult females in the sealed cell do not produce eggs, especially those on female broods (9). The research carried out by Avdeeva (2) and confirmed by Hänel (8) showed the key role played by the juvenile hormone JH_{III} in inducing egg laying.

Table II - Survival of mites in different artificial environments and in a natural comb.

Environment	Initial No. of bee larvae	adult bees obtained	surviving mites
ELISA plate	37	36	31
WAX plate			
with worker larvae	50	46	40
with drone larvae	40	12	11
Piece of an artificially infested comb	33	33	33
Piece of a naturally infested comb	54	53	46

Table III - Adult females of **Varroa jacobsoni** surviving after each of 4 consecutive transfers on worker bee larvae.

day	0	12	21	27	34
surviving mites	30	24	19	13	9

The possibility was also checked that the low egg production under artificial conditions was due to other factors such as disturbance after being moved and the variations in temperature.

With this aim mites were transferred from cells, which had been sealed for less than two days, and thus had not yet produced Varroa eggs, to other larvae in the same comb and of approximately the same age, through a small hole in the operculum which was immediately closed. As a control two pieces of naturally infested comb, one adjacent to that to which the transfer had been made and so also subjected to room temperatures for the three hours necessary for the transfer, and the other kept all the time in the incubator. The cells were examined when pupae were at the stage of dark eyes and egg laying had mostly taken place. The presence or absence of eggs was noted in each cell together with the number of progeny. Cells which in addition to the artificially introduced mite also contained another mite, due to a previous natural infestation, were discarded, as well as those in some way damaged or with an abnormally developed host. The results are shown in Table IV.

The number of mites which reproduce and the number of progeny is slightly higher in those taken from drone brood; it does not appear that the transfer itself strongly influences the ability of the mites to reproduce. However the data do suggest that exposure to relatively low temperatures (about 20°C) for a few hours can reduce fecundity, as was observed (1) for periods of 5-6 hours. In the end it must be pointed out that the number of mites which reproduce and the number of progeny produced in naturally infested combs and kept in the incubator are comparable to those found in nature (10).

4.Conclusions

This preliminary investigation emphasized certain interesting problems connected with the breeding of **Varroa jacobsoni** outside its natural environment.

Table IV - Number of females of **Varroa jacobsoni** which lay eggs and number of the progeny in pieces of comb kept in the incubator.

	No. of mites	Ovipositing females	Total No. of progeny
A) Infested with mites taken from drone brood	25	14	28
B) Infested with mites taken from worker brood	24	11	19
C) Natural infestation	24	12	18
D) Natural infestation and immediate transfer to the incubator	24	18	54

In regard to survival the greatest threat comes from the development
of moulds, perhaps on injuries of the host, and which would probably be
controlled inside the hive by natural antagonists about which we have
little information. It has already been noted (2) that **in vitro** these are
an important cause of death. Concerning reproduction the obstacles
encountered by other investigators were confirmed by us. A regular rhythm
of egg laying and not just sporadic development of immature stages could
not be achieved. This is due both to the difficulty of producing the key
stimuli which, synchronizing the life cycle of the parasite with that of
the host, allows the former to survive and both to the sensitivity of the
mite to all the variations in the physical-chemical conditions of its
environment.

A very useful technique is described for transferring mature larvae
from combs. This work can be done rapidly and with a high rate of success.

Finally the aggregation of **Varroa jacobsoni** on bee larvae was observed. This will be further investigated in nature to find the quantitative aspects and to clarify the determining factor, as the phenomenon may
be due to the existence of aggregating substances, released by the mite,
of potential use in the diagnosis and in the control of the pest.

References

1. AKIMOV, I. A., PILETSKAIA, I.V., (1985). [The effect of temperature on
   the deposition and development of eggs of **Varroa jacobsoni**.] Vestnik
   Zoologii No. 3, 52-56. (In Russian).
2. AVDEEVA, O.I. (1978). [Life cicle of the **Varroa** mite in laboratory
   conditions.] Pchelovodstvo 1978, No. 10, 16-17. (In Russian).
3. AVDEEVA, O.I. (1979). [The biology of nutrition of the **Varroa** mite
   under laboratory conditions]. Pchelovodstvo 1979, No. 8, 18-19. (In
   Russian).
4. BUEDEL, A. (1961). Le microclimat de la ruche, in: R. CHAUVIN, Traité
   de la biologie de l'abeille. Masson et Cie, Paris.
5. CAVICHO ISSA, M.R. (1983). A technique for inducing egg laying by
   Varroa jacobsoni in the laboratory. In: Apimondia XXIX, Bucharest.
6. CAVICHO ISSA, M.R., GONCALVES, S. (1985). La technique de l'induction
   de la ponte chez l'acarien **Varroa jacobsoni** en conditions de
   laboratoire. In: Apimondia XXX, Bucharest.
7. DELFINADO-BAKER, M. (1984). The nymphal stages and male of **Varroa
   jacobsoni** Oudemans a parasite of honey bees. Internat. J. Acarol., **10**,
   75-80.
8. HAENEL, H. (1983). Effect of JH$_{III}$ on the reproduction of **Varroa
   jacobsoni**. Apidol. **14**, 137-142.
9. IFANTIDIS, M.D. (1983). Ontogenesis of the mite **Varroa jacobsoni** in
   worker and drone honeybee brood cells. J. Apic. Res., **22**, 200-206.
10. IFANTIDIS, M.D. (1984). Parameters of the population dynamics of the
    Varroa mite on honeybees. J. Apic. Res., **23**, 227-233.
11. SAKAI, T.; TAKEUCHI, K.; HARA, A. (1979). [Studies on the life history
    of a honeybee mite, **Varroa jacobsoni** Oudemans, in laboratory rearing.]
    Bull.Fac. Agric., Tamagawa Univ., 1979, No. 19, 95-103. (In Japanese).

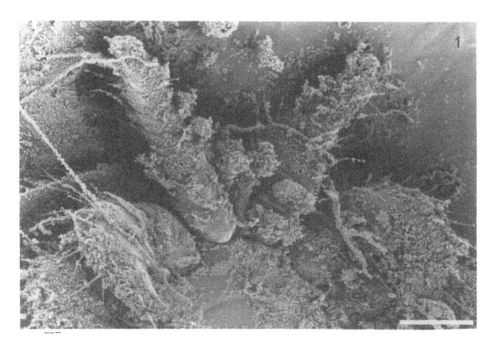

Fig. 1 - Body, legs and mouth parts of a living **Varroa jacobsoni**, reared on bee larva in the laboratory, covered with fungal conidia. Scale bar: 200 μm.

Fig. 2 - Detail from the previous picture. Scale bar: 100 μm.

Fig. 3 - **Varroa jacobsoni** palps bearing masses of fungal conidia. Scale bar: 40 μm.

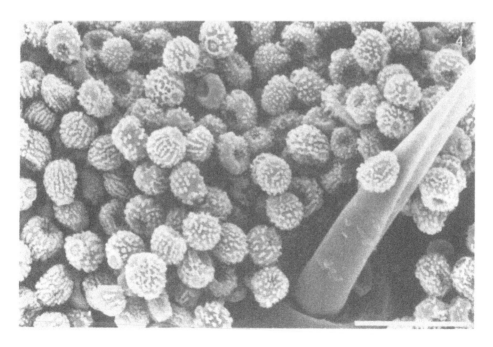

Fig. 4 - Detail from the previous picture. Scale bar: 10 μm.

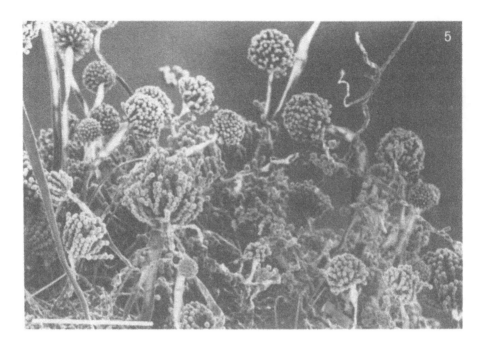

Fig. 5 - **Aspergillus** sp. conidiophores on a dead **Varroa jacobsoni** reared
in the laboratory on bee larva. Scale bar: 100 μm.

Fig. 6 - Detail from the previous picture. Scale bar: 10 μm.

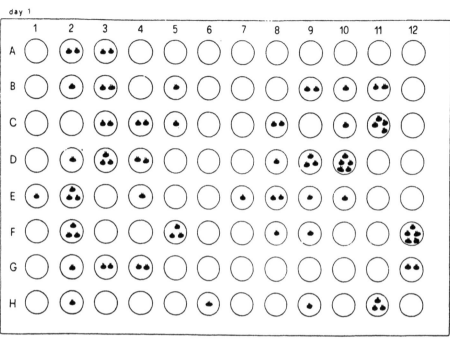

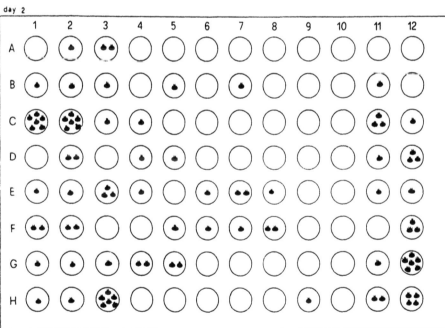

Fig. 7,8 – Distribution of **Varroa jacobsoni** on worker bee larvae in a micro-ELISA plate during the days 1-2 following the infestation.

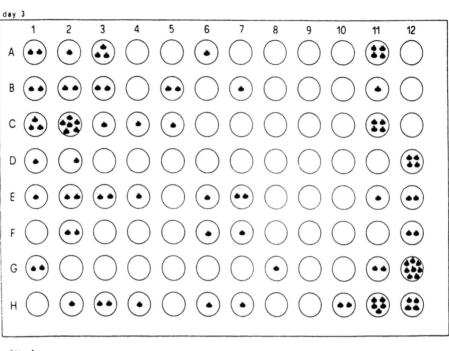

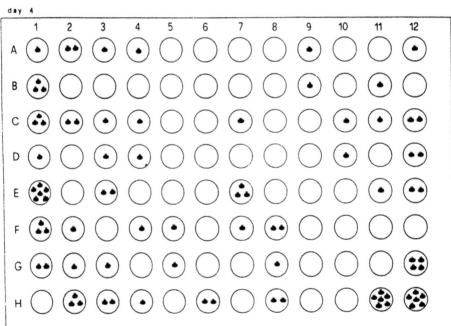

Fig. 9,10 - Distribution of **Varroa jacobsoni** on worker bee larvae in a micro-ELISA plate during the days 3-4 following the infestation.

124

An artificial diet for *Varroa jacobsoni* Oudemans (Acari: Varroidae)

W.A.Bruce
Istituto di Difesa delle Piante, Università degli Studi, Udine, Italy, and Stored-Product Insects Research and Development Laboratory, USDA, Agricultural Research Service, Savannah, Ga., USA
F.Chiesa
Istituto di Difesa delle Piante, Università degli Studi, Udine, Italy

Summary

Within the enforced constraints of a four month study it has been demonstrated that:
(I) All feeding stages of **Varroa jacobsoni** are able to feed upon an artificial diet obtained through a synthetic membrane.
(II) Females oviposited within the environment of an artificial cell.
(III) Artificial diet was seen to be incorporated within eggs.
(IV) Eggs hatched but protonymphs failed to complete development to the next stage, thus a complete life-cycle has yet to be realised.
Further studies to achieve the above goal will be carried out at the University of Udine, Institute for Plant Protection. A more complete account of the work described in this condensed version will be published in a separate communication.

1. Introduction

Varroa jacobsoni Oudemans is a mite ectoparasite of honeybees and is found throughout Europe and Asia. Its potentially devastating impact on apiculture should neither be ignored nor underestimated in those countries where infestations already occur. Countries such as the United States, Canada, Australia, and New Zealand fortunate enough to be free from infestation must not become complacent but must partecipate actively in cooperative efforts for the study, containment, and elimination of this serious threat to world agriculture. In this regard, the European Economic Community (EEC) countries have taken the lead in a coordinated research effort to study both the basic and applied aspects of this problem.

Because of the unique biology of this parasite, adequate study of its behaviour has been greatly impaired. It is extremely difficult to study complexities of feeding, mating, and oviposition behaviour when the organism is sealed in an opaque cell sequestered in a darkened beehive.

It comes as no great surprise then that researchers worldwide would like to have made available to them the techniques necessary to produce large numbers of standardized, laboratory-reared mites of all stages. The possibility of rearing **V. jacobsoni** in the laboratory on an artificial diet was suggested by Dr. Donald Griffiths (personal communication)

125

following the Third International Working Conference on Stored-Product Protection, Manhattan, Kansas, USA, 1983. The presentation that one of us (W. A. B.) had given at this meeting concerned the laboratory rearing of **Pyemotes tritici,** an ectoparasite of stored-product insects. While both **P. tritici** and **V. jacobsoni** pierce the cuticle of their host insect and feed on its haemolymph, further comparison of their biologies is difficult. However, speculation as to the possibility that **V. jacobsoni** would also feed on the same artificial medium as that developed for **P. tritici** and would do so through the same synthetic membrane seemed to us probable and worthy of study. These discussions resulted in a proposal being submitted to and accepted by the EEC, Experts' Group on Varroatosis. The overall objective was clear: develop an artificial rearing technique for **V. jacobsoni.** The research was conducted at the Institute for Plant Protection, University of Udine, Italy, during a 4 month period (May 1 - August 31, 1986). This paper summarizes the results of that research.

The specific objectives of this study were: (1) Determine if adult female **V. jacobsoni** would probe, pierce and feed through a non-living, synthetic membrane; (2) Determine, provided a suitable membrane was found, if the other life stages would also feed through the membrane; (3) Determine if **V. jacobsoni** would feed on a non-insect, artificial diet and lay eggs; (4) Determine if **V. jacobsoni** could complete its life cycle fed an artificial diet and reared under laboratory conditions.

2. Materials and methods

All life stages of **V. jacobsoni** were obtained from hives of the domestic honeybee **Apis mellifera.** Artificial cells which simulated those of the natural comb were prepared from suitably modified plastic queen cells or from a micro-ELISA test plate. A number of synthetic membranes was selected and tested. Initial tests to determine the ability of the mite to feed through the membrane employed filtered homogenate of larval and pupal haemolymph, dyed with methylene blue. Subsequently, an artificial diet (Bruce, unpublished) was prepared and tested under the same conditions.

3. Results and discussion

The synthetic membrane eventually selected as most suitable was Parafilm, all others tested gave negative results. Parafilm drawn out to a thickness of approximately 10 μm proved to be highly effective.

Initial tests using the modified ELISA plate and dyed homogenised haemolymph showed, from the presence of coloured faeces, the ability of female Varroa to feed through the membrane. However, the ELISA test plate was abandoned due to loss of diet through leakage problems.

Thus, modified queen-rearing cells were adapted for the remainder of the tests and artificial diet was substituted for homogenised haemolymph.

Using this technique females fed through the membrane and laid eggs upon its surface or on the sides of the cell, entirely under laboratory conditions (34±1 °C and 85% RH).

Evidence of methylene blue within some eggs indicated egg development was directly related to the ingestion of artificial diet. About half of those eggs laid hatched to the protonymphal stage. Of these, some were observed to feed but none moulted to the deutonymphal stage. In turn, deutonymphs taken directly from brood cells fed, but again failed to develop further. It should also be noted that females offered a free choice of feeding naturally upon a bee pupa or through the artificial membrane fed randomly on both, as evident through the presence of white and blue faeces within the test cell.

Session 4
Control methods

Chairmen: R.Borneck
O.Van Laere
F.Frilli

Bromopropylate decay and residues in honey samples

M.Barbina Taccheo & M.De Paoli
Sezione inquinamento agrario e difesa biologica dell'ambiente, Centro Regionale per la Sperimentazione Agraria per il Friuli-Venezia Giulia, Pozzuolo del Friuli, Italy

S.Marchetti
Istituto di Produzione Vegetale, Università degli Studi, Udine, Italy

M.D'Agaro
Istituto di Difesa delle Piante, Università degli Studi, Udine, Italy

Summary

Results of a three year research study on the determination of bromopropylate residues in honey samples are reported. Analyses were made by capillary gas-liquid chromatography with electron-capture detection. The samples of honey analyzed were collected from hives at different intervals following treatment. Additional samples from other treated hives were kept at room temperature to verify the decay of the active ingredient without the dilution by uncontaminated honey that normally would have been produced in the hive following treatment. The effect on the amount of residues of i) the position of the fumigant strip during treatment, and ii) the level of sealing in the combs filled with honey was investigated. The distribution of the a.i. in combs that occupied fixed positions inside the hive was studied for two methods of fumigation and two levels of cell sealing. The trial enabled to ascertain the remarkable persistance of bromopropylate in the honey matrix, especially when honey had been directly exposed to smoke during treatment. The 'dilution' of the a.i. with the newly imported honey, and the honey consumption from bees gave no warranty against the presence of residues in the honey crop.

1. Introduction

The possible presence of pesticide residues in honey was considered for the first time only at the end of '60s. Since then, the research was particularly significant in Italy, West Germany and some countries of Eastern Europe. The work (1, 3, 5, 6, 7, 8, 9, 10, 11, 13, 14, 15, 16, 18, 21, 22, 23, 24, 25, 28, 30, 31, 32, 33, 34, 36, 37, 38, 39) dealt and deals with indirect contamination of honey due to the application of insecticides and herbicides to agricultural crops.

The introduction of the parasitic mite Varroa jacobsoni Oud. caused a further boost to research in this field (4, 12, 19, 29, 35); this situation will probably continue as chemical treatment will presumably keep on being more effective and less time-consuming than biological methods of control.

Fig. 1 – Application of the fumigant strip in an empty honey chamber.

Fig. 2 – Application of the fumigant strip inside the nest.

132

A method for the determination of bromopropylate and two of its degradation products in honey was developed by Formica (12). Later, Barbina Taccheo and coll. (4) described a multiresidue method for the determination of bromopropylate, tetradifon, and malathion in honey samples. The present work shows the results obtained when applying this method over a quantity of honey samples collected from colonies treated with bromopropylate; this chemical is contained in the trade product Folbex VA (Ciba-Geigy) which at this moment is still the only product registered in Italy for the control of varroatosis. The trials were to check on the persistance of the substance in honey samples so as to define a safe interval between treatment of the colonies and honey consumption. The samples of honey analysed were collected from hives at different intervals following treatment. Additional samples from other treated hives were kept at room temperature to verify the decay of the active ingredient without the dilution by uncontaminated honey that would normally have taken place in the hive following treatment. The effect of the following factors on the amount of residues was also investigated: i) position of the fumigant strip during treatment; ii) level of sealing in the combs filled with honey. Finally, the distribution of the a.i. in combs that occupied certain fixed positions inside the hives was studied for two methods of fumigation and two levels of cell sealing.

2. Materials and Methods

None of the hives used had been treated with bromopropylate before these trials. In the first 3 tests the fumigant strip was placed inside the nest; the exact position during treatment is indicated in Fig. 2. This is the most common way of using Folbex VA in the Friuli Region (north east of Italy).

The first test followed a trial in which the effectiveness of Folbex VA as well as other acaricide products was studied (20). Three hives were treated only four times at four day intervals (date of the first fumigation: October 18, 1983). Honey samples were collected 60, 135, 180, 205, and 232 days after the fourth treatment. These were taken from the upper part of the nest combs, where cells were capped. A small spatula was driven into each comb and then turned to facilitate the exit of the honey. Each sample consisted of 40 g of honey. On March 27, 1984 the veterinary services applied an additional treatment with Folbex VA, without the honey supers added. On May 22, 1984 that is 205 days after the last autumn treatment and 56 days after the last fumigation, honey samples were collected also from the honey supers.

The second test was carried out on 4 colonies treated in the same way as the previous ones. The first treatment with Folbex VA was applied on October 16, 1984. Honey samples were extracted 2, 23, 48, 201, 227, 261, 297, and 324 days after the fourth fumigation. In contrast to the first test, we tried to collect, from each colony, a honey sample representative of the whole honey content. The spatula was inserted into more than 40 positions vs. 10 of the former test. On August 21, 1985 we collected

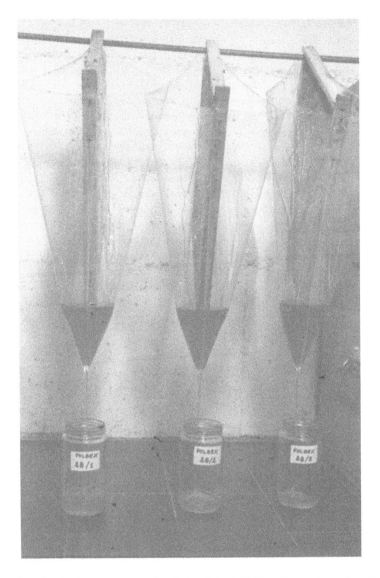

Fig. 3 - Sample technique used in the third test; jars were stored at 25
°C in a climatic chamber.

samples from the honey chambers.

The third test was carried out in a similar way but on three colonies
and the date of the first fumigation was March 25, 1985. Honey was sampled
9, 43, 139, and 166 days after the fourth fumigation. Samples from honey
supers were collected on the same date as in the former test.

On June 25, 1986 twelve hives received the first treatment with
Folbex VA. The number of treatments and the interval between two of them
remained unvaried. In 6 hives the fumigant strip was hung in an empty

honey chamber (Fig. 1) as suggested by the manufacturers; in the remaining 6 the strip was placed inside the nest as in the treatments carried out in the previous years (Fig. 2). Honey was extracted only from the combs indicated in Figs. 1 and 2. In 6 hives (3 for each method of fumigation), combs with honey were almost completely sealed at the beginning of the trial. In the remaining hives caps were removed with an uncapping fork before each treatment. When treatments were completed, honey contained in single combs was extracted by centrifugation (caps in the sealed combs were removed with an uncapping fork). To avoid mixtures of honey in the honey-extractor, each comb was put in a large plastic bag from which a kilogramme of honey was collected and kept at 25 °C in a climatic chamber (Fig. 3). Honey was sampled and analysed 3, 10, 17, 24, 31, 45, 59, and 73 days after the fourth treatment.

Further analyses were carried out on samples of caps present inside colonies during treatment. Here the method of analysis was slightly changed; the samples (5 g of wax each) were extracted twice with a 10 ml solution of n-hexane/ethyl ether (50 : 50), filtered on adsorbent paper and dil to 25 ml with the same solvent solution.

In order to assess the influence of wax particles on the amount of residues, two samples of unfiltered honey taken from treated colonies were tested. Caps were isolated by filtration and analyzed separately. Each honey sample was poured into a 20 ml syringe; these were put in a lab centrifuge at 6000 r.p.m. for 15 min. From one syringe we collected honey at the bottom and at the top, from the other at the bottom and at the middle. Honey was analyzed according to the method used in the former tests.

3. Results and Discussion

Results of chemical analyses carried out on honey samples collected directly from the hives are shown in Tab. I; they indicate that bromopropylate is a persistance substance in the honey matrix; 324 days after the last treatment with Folbex VA, detectable amounts of bromopropylate were still present within the nest combs. From data reported, it can also be deduced that beekeepers should not rely either on the 'dilution' of the a.i. with honey produced in the months following treatment, or on the consumption from the honeybee colonies during winter or spring. Indeed, the second factor can even lead to an increase in the concentration of the a.i. at the beginning of spring (cf. Tab. I, second test); this may happen when lateral combs (i.e. those nearer to the fumigant strip) are not moved towards the center of the nest so stimulating bees to feed on the most contaminated honey.

Samples taken from the honey chambers also contained bromopropylate (Tab. I), although treatment was carried out without the honey supers added and long before the nectar flows. The rule of applying treatment at least 15 days before the addition of the honey supers was found to give no warrant against the presence of residues in the honey crop.

DATE OF THE FIRST TREATMENT	DAYS AFTER THE 4TH TREATMENT	RESIDUES (PPM)	
		NEST	SUPER
OCT. 18, 1983	60	.101	
	135	.031	
	180 (31) (*)	.199	
	205 (56)	.186	.095
	232 (83)	.070	
OCT. 16, 1984	2	.125	
	23	.117	
	48	.097	
	129	.085	
	201	.098	
	227	.043	
	261	.042	
	297	.068	.034
	324	.063	
MARCH 25, 1985	9	.226	
	43	.167	
	139	.115	.044
	166	.069	

(*): Days after the additional treatment carried out by veterinary services.

Tab. I - Bromopropylate residues (ppm) found at different intervals following treatment when honey samples were taken from the hives.

	RESIDUES (PPM)
FIRST SAMPLE	
HONEY COLLECTED	
AT THE TOP	.203
AT THE BOTTOM	.100
CAPS	12.368
SECOND SAMPLE	
HONEY COLLECTED	
AT THE MIDDLE	.156
AT THE BOTTOM	.161
CAPS	11.500

Tab. II - Bromopropylate residues (ppm) found in caps and in honey taken from different positions of the syringes.

We must remember that the sale of honey with any degree of chemical contamination is absolutely forbidden in Italy. More precisely, up till June 5, 1985 any residue was not permitted (26); but having considered possible indirect contamination, the Health Ministry decided to consider as absent a substance less than 0.01 ppm concentrated in foods (27).

The persistance of bromopropylate in honey was confirmed by data regarding the decay of the a.i. at room temperature. In Fig. 4 we present 4 curves from the 36 available. These curves (as well as all the others) show that, from a mathematical point of view, the amount of residues determined a few days after the 4th treatment with Folbex VA can be easily related to those found in the following weeks; actually, in the analysis of regression according to an exponential model we always observed levels of significance at least equal to 5 %. In addition, we noted that the trends described by different equations often differed among themselves, in the sense that sometimes the amount of residues found about 10 weeks after the 4th fumigation could be the same even if wide differences had been observed in the first samplings. We also noted that four weeks after the end of treatment only slight variations could be detected in the amount of residues. It should be remembered that each curve was derived from a different large sample and that each sample was collected from a different comb. The reason which determined the difference in the rate of decay was discovered and will be described later.

In Fig. 4 it can be observed that sometimes an increase in residues between two successive analyses may be recorded. Two reasons are possible: instrumental errors or the presence in the honey of some substance in which bromopropylate is more soluble. Chemical properties of the a.i. (2) suggested our verifying whether this interferent was the wax. Results of analyses carried out on honey collected at different positions within the syringes after centrifugation, and on the two cap samples are shown in Tab. II. It should be emphasized that honey was filtered using the same method as beekeepers. After centrifugation, honey at the top of the syringe visually differed from the rest; more precisely, a thin brownish layer could be distinguished. Results achieved from 5 other cap samples agreed with those reported in Tab. II as amounts of residues were about 100 times greater than those found in honey extracted from the relative cells.

Since it is not feasible to modify the filtration process so as to decrease the level of honey contamination, we tried to draw useful information from data regarding the amounts of residues obtained with two methods of fumigation in nest combs having a different position during treatment (Figs 1 and 2). These results are shown in Tab. III. Honey extracted from comb no. 2 was found by far to be the most contaminated when the fumigant strip was hung inside the nest. With the other method of fumigation no appreciable differences among combs were observed. The level of contamination was higher when treatments were applied in the nest and reached its maximum in case of association between this method of fumigation and the presence of unsealed combs. Surprisingly, honey contained in sealed combs was not immune from contamination. However, the

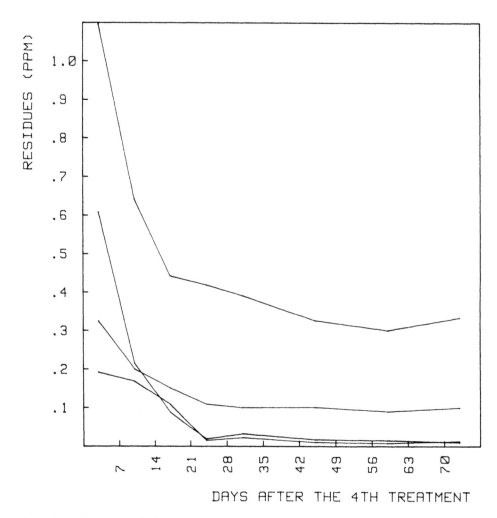

Fig. 4 - Bromopropylate decay in 4 honey samples (residues are expressed in ppm).

factor which had the most influence on the final amount of residues was still the presence of caps on the cells filled with honey. In fact, the bromopropylate decay was found to be much higher in honey extracted from sealed combs than in honey directly exposed to smoke during treatment. Seventy-three days after the last fumigation, all the samples contained a residue concentration higher than that permitted.

4. Conclusions

The trials showed the remarkable persistance of bromopropylate inside the honey matrix. The 'dilution' of the a.i. with the newly imported honey, and the honey consumption from bees gave no real warranty against

138

DAYS AFTER THE 4TH FUMIGATION

		3	10	17	24	31	45	59	73
FUMIGANT STRIP INSIDE THE NEST									
HONEY FROM:									
SEALED COMBS	1	.160	.094	.072	.034	.033	.025	.023	.027
	2 (*)	.560	.301	.194	.158	.165	.159	.176	.179
	3	.204	.146	.098	.064	.072	.072	.062	.057
	AVERAGE	.308	.180	.121	.085	.090	.085	.087	.088
UNSEALED COMBS	1	.396	.343	.253	.201	.206	.196	.178	.184
	2	.715	.515	.367	.321	.300	.276	.263	.272
	3	.418	.323	.188	.161	.150	.149	.147	.151
	AVERAGE	.509	.394	.269	.228	.219	.207	.196	.202
FUMIGANT STRIP INSIDE AN EMPTY SUPER									
HONEY FROM:									
SEALED COMBS	1	.275	.108	.056	.028	.036	.030	.024	.023
	2	.298	.148	.080	.075	.074	.078	.070	.073
	3	.279	.173	.082	.067	.078	.076	.071	.068
	AVERAGE	.284	.143	.073	.057	.063	.061	.055	.055
UNSEALED COMBS	1	.309	.217	.191	.127	.137	.130	.132	.123
	2	.247	.205	.174	.147	.150	.141	.143	.144
	3	.296	.293	.185	.147	.137	.126	.126	.120
	AVERAGE	.284	.238	.183	.140	.141	.132	.134	.129

(*): in one replicate (i.e. comb) only 70 % of the total surface was sealed before the 1st treatment vs. 90-100 % of all the other combs classified as sealed

Tab. III - Bromopropylate residues (ppm) found at different intervals following treatment in sealed and unsealed combs occupying fixed positions inside the hive (see also Figs. 1 and 2).

the presence of residues within the honey crop. To reduce or avoid contamination, the best solution would be to remove the combs filled with honey from the nest and keep them in a store during the treatment period. If these operations are judged laborious and time-consuming, it would be opportune to adopt at least the following precautions: 1) apply treatments only during the autumn season; 2) mark the unsealed combs with honey; 3) hang the fumigant strip between two frames in the honey super; otherwise, 4) hang the fumigant strip inside the nest between a wooden frame (or the last comb which contained brood) and a wooden screen, leaving the combs filled with honey as far as possible beyond the screen; 5) in the following spring uncap and move combs towards the center of the nest so as to stimulate bees to feed on the contaminated honey; 6) feed the bees with sugar syrup only when strictly necessary; 7) be sure to filter the honey well; 8) do not sell honey extracted from the nest.

References

1. ANON. (1972). Results of food control in Switzerland in 1971. Mitteilungen aus dem Gebiete der Lebensmitteluntersuchung und Hygiene, 63 (3), 321-423.

2. ANON. (1983). The Pesticide Manual. 7th edition, The British Crop Protection Council, pp. 695.

3. BALDI M., BOVOLENTA A., PENAZZI L., ZANONI L. (1981). La contaminazione degli alimenti da pesticidi clorurati e fosforati: due anni di ricerche (1979-1980) su alimenti prodotti nella provincia di Ferrara. Industrie alimentari, 20 (10), 691-700.

4. BARBINA TACCHEO M., DE PAOLI M., SPESSOTTO C., MARCHETTI S. (1985). Metodo multiresiduo per la determinazione nel miele di alcuni principi attivi impiegati contro Varroa jacobsoni Oud.. Atti 5° Simposio Chimica degli antiparassitari - Residui, metaboliti e prodotti di trasformazione, Piacenza, 6-7 Giugno 1985, 115-123.

5. BENTLER W., FRESE E. (1981). Microbial quality and residue analysis of honey. Archiv fuer Lebensmittel-Hygiene, 32 (4), 130-135.

6. BLAHA J. J., JACKSON P. J. (1985). Multiresidue method for quantitative determination of organophosphorus pesticides in foods. J. Assoc. Anal. Chem., 68 (6), 1095-1099.

7. BIGAZZI GRASSO C., CAPEI R. (1983). Sulla presenza dei residui di insetticidi clorurati nel miele. L'Igiene Moderna, 80 (5), 975-980.

8. BRUNS G. W., CURRIE R. A. (1983). Determination of 2-chloroethanol in honey, beeswax, and pollen. J. Assoc. Off. Anal. Chem., 66 (3), 659-662.

9. CHERKASOVA A. I., ISTRATOV I. F. (1973). Dynamics of pesticide residues in apicultural products. Bdzhil'nitstvo, 9, 79-81.

10. CERUTTI G., MANNINO S. (1979). Ferro, rame, manganese e residui cloroorganici nel miele. La difesa antiparassitaria nelle industrie alimentari e la protezione degli alimenti, 1, 489-492.

11. DANIELYAN S. G., NALBANDYAN K. M., MARKOSYAN A. A. (1976). Determination of residual quantities of organophosphorus toxic materials and sevin in bees, honey, and beebread. Izv. S-kh. Nauk, 19 (4), 53-57.

12. FORMICA G. (1984). Gas cromatographic determination of residues of bromopropylate and two of its degradation products in honey. J. Assoc. Off. Anal. Chem., 67 (5), 896-898.

13. GAYGER J., DUSTMANN J. H. (1985). Investigations for residues of pesticides in the honeybee products beeswax, honey, and pollen. Archiv fuer Lebensmittel-Hygiene, 36 (4), 93-96.

14. GRANDI A. (1975). Preliminary research on contents of chlorinated insecticides and phosphoric esters in Italian honeys. Scienza e tecnologia degli alimenti, 5 (2), 103-105.

15. HAUBRY J. (1975). Various centres for hollow shapes. Revue des fabricants de confiserie, chocolaterie, confiturerie, biscuiterie, 50 (1), 25-30, -(2), 24-28.

16. JUMAR A., SIEBER K. (1967). Residue studies in rapeseed oil and honey with toxaphene-chlorine-30. Z. Lebensm.-Unters. Forsch., 133 (6), 357-364.

17. LEONI V., MARCHESE E. (1985). La nuova ordinanza sulle quantità massime di antiparassitari consentite negli alimenti. Atti 5° Simposio Chimica degli antiparassitari - Residui, metaboliti e prodotti di trasformazione, Piacenza, 6-7 Giugno 1985, 7-17.

18. LESHCHEV V. V., SMIRNOV A. M., TALANOV G. A. (1974). Determination of the insecticide phosalone in honey by TLC and GLC methods. Veterinariya, Moscow, no. 6, 106-107.

19. MALININ O. A., YAROSHENKO V. I., ALEKSEENKO F. M. (1981). Determination of neoron in honey. Veterinariya, Moscow, no. 9, 68-69.

20. MARCHETTI S., BARBATTINI R. (1984). Comparative effectiveness of some treatments used to control Varroa jacobsoni Oud.. Apidologie, 15 (4), 363-378.

21. MCLEOD H. A., MENDOZA C. E., MCCOLLY K. A. (1975). Analysis of foods for methomyl using thin-layer chromatography after low temperature clean-up. Pestic. Sci., 6 (1), 11-16.

22. MESTRES R., ESPINOZA C., CHEVALLIER C., MARTI G. (1979). Determination of decamethrin residues. Travaux de la Societe de Pharmacie de Montpellier, 39 (4), 329-336.

23. MUELLER B. (1973). Determination of pesticide residues in bee honey. I. Semiquantitative thin-layer chromatographic determination of insecticide residues in bee honey. Nahrung, 17 (3), 381-386.

24. MUELLER B. (1973). Determination of pesticide residues in bee honey. II. Semiquantitative thin-layer chromatographic determination of herbicide residues in bee honey. Nahrung, 17 (3), 387-392.

25. OGATA J. N., BEVENUE A. (1973). Chlorinated pesticide residues in honey. Bulletin of environmental contamination and toxicology, 9 (3), 143-147.

26. ORD. MIN. (6/1/1979). Gazzetta Ufficiale, no. 39, 8/2/1979.

27. ORD. MIN. (6/6/1985). Gazzetta Ufficiale, no. 250, 23/10/1985.

28. PETUKHOV R. D. (1976). Determination of anthio and phosphamide in honey. Veterinariya, Moscow, no. 7, 101-102.

29. PETUKHOV R. D. (1981). Determination of Kelthane in honey by thin-layer chromatography. Byulleten' Vsesoyuznogo Instituta Eksperimental'noi Veterinarii, 41, 74-76.

30. POURTALLIER J., TALIERCIO Y. (1967). Toxic residues in honeys. II. Detection and estimation of pesticide residues. Bull. Apic. Inform. Doc. Sci. Tech., 10 (1), 51-62.

31. RIPEANU M., CRAINICEANU E., TANASE C. (1968). Detection of nitrogen-containing organophosphorus compounds in dead bees and apicultural products. Apicultura (Bucharest), 21 (3), 16-18.

32. SABATINI A. G., SAVIGNI G. (1976). Residues of organochlorine and organophosphorus pesticides in samples of honey from Emilia-Romagna. Rivista di scienza e tecnologia degli alimenti e di nutrizione umana, 6 (3), 167-170.

33. THIELEMANN H., GRAHNEIS H. (1978). Semiquantitative TLC determination of toxaphene in honey - A contribution to residue analysis. Zeitschrift fuer die Gesamte Hygiene und Ihre Grenzgebiete, 24 (2), 139.

34. THIELEMANN H., GRAHNEIS H. (1983). Experimental investigations on the quantitative thin-layer chromatographic determination of toxaphene by evaluation of the chromatograms using an extinction registering device with integrator "ERI 65 m" (VEB Carl Zeiss Jena). Zeitschrift fuer die Gesamte Hygiene und Ihre Grenzgebiete, 29 (7), 374-375.

35. THRASYVOULOU A. T., IFANTIDIS M. D., PAPPAS N. L., SIMMONS K. (1985). Malathion residues in greek honey. Apidologie, 16 (1), 89-94.

36. TSVETKOVA T., PENEVA V., GRIGOROVA D. (1981). Residual pesticides in honey. Veterinarnomeditsinski Nauki, 18 (1), 93-98.

37. VAS'KOVSKAYA L. F., KLISENKO M. A. (1968). Determination of Tedion residues in honey by thin-layer chromatography. Khim. Sel. Khoz., 6 (4), 285-286.

38. WESTOEOE G., NOREN K. (1973). Residues of organochlorine pesticides and certain organophosphorus pesticides in fruits, berries, vegetables, and roots, 1968-June 1972. Var Foeda, 25 (1 suppl.), pp. 44.

39. YAROSHENKO V. YI. (1974). Dosage du chlorophos dans la miel par methode enzymatique de diffusion dans la gelose. Visn. Sil'skogospodar. Nauki, no. 1, 108-111.

Reinfestation rates of Varroatosis after treatments in brood-free honeybee colonies

S.Marchetti
Istituto di Produzione Vegetale, Università degli Studi, Udine, Italy
F.Frilli & M.D'Agaro
Istituto di Difesa delle Piante, Università degli Studi, Udine, Italy

Summary

The trial was conducted during the active season of 1984 on 80 colonies. Every month, in a set of 10 colonies the queens were confined in cages and four treatments with Folbex VA were carried out in absence of brood. The number of mites that survived the treatment was related to the number of mites found in October. The effect on the entity of reinfestation of a varying rate of the mite reproduction and mortality throughout the active season was studied. The trial enabled to ascertain a remarkable rate of reinfestation in the treated hives so as to recommend annual treatment, even when the fumigations are carried out in the complete absence of brood.

1. Introduction

To get the best results in the chemical control of varroatosis it is necessary to treat colonies with little or no brood. These conditions are easily met in most environments of northern Italy during the cold seasons. Conversely, in warmer regions such as Sicily the amount of brood is considerable even in January, therefore the only brood-free colonies are normally those which have given rise to swarms and of course the swarms themselves.

Methods for getting brood-free colonies have been proposed (13, 7, 11, 16, 17, 20, 5, 12, 10, 4) to enable beekeepers to carry out the biotechnical control of varroatosis or to enhance the effects of chemical treatments. Much of the applicability of these methods depends on the required amount of extra work and on whether and how this extra work is related to the normal or generally desirable management of bees.

Treating bees of brood-free nuclei and of brood-free mother colonies could be 'considered a good technique; maybe using the simplest experimental method to get brood-free colonies, i.e. caging the queens for 21 to 24 days, could also be profitable. Before recommending these techniques it would be of great help to know whether they must be applied yearly or at longer intervals and which breakdown in the parasite populations must be induced to avoid any danger of recrudescence of the disease for the suggested pattern of control.

145

An exhaustive answer to these problems involves trials lasting at least two years. The present one-year research study should therefore be regarded as a first attempt to learn something more about the rates of reinfestation in colonies where a noticeable reduction in the mite number was reached by means of treatments in the absence of brood.

2. Materials and Methods

The trial was conducted on the same hives used to assess the effectiveness of the following diagnostic techniques: i) chemical treatment in presence of brood; ii) examination of an insert kept on the bottom board for a 15 day period; iii) examination of a sample of adult bees; iiii) examination of a sample of capped drone brood. Data concerning the effectiveness in the diagnosis were collected each month, from April to September of 1984. We worked on groups of 10 colonies per month according to the following schedule:

```
                    EI,AB,DB,QC,F1
 CI                                      F2                        F3  F4  F5  F6
I---I---I---I-/\/-I---I---I---I---I---I---I---I---I-/\/-I---I---I---I
15  16  17  18    1   2   3   4   5   6   7   8   9  25  26  27  28
```

CI= change of the insert;
EI= extraction of the insert;
AB = drawing of a sample of 500 adult bees;
DB = drawing of a sample of capped drone brood (if present);
QC = caging of the queen;
F1 = 1st treatment with Folbex VA;
F2 = 2nd treatment with Folbex VA;
F3 = 1st treatment with Folbex VA in absence of brood;
F4 = 2nd treatment with Folbex VA in absence of brood;
F5 = 3rd treatment with Folbex VA in absence of brood;
F6 = 4th treatment with Folbex VA in absence of brood.

In autumn all the available colonies were treated again using bromopropylate (trade name: Folbex VA, Ciba-Geigy). Treatment was applied as follows:

```
  ▼   ▼           ▼   ▼           ▼   ▼           ▼   ▼
I---I---I---I---I---I---I---I---I---I---I---I---I---I---I---I---I
 1   2   3   4   5   6   7   8   9  10  11  12  13  14  15  16  17
```

At least 4 fumigations were carried out in the absence of brood, as rearing had stopped naturally ahead of time.

Former research, carried out by the Istituto di Difesa delle Piante, University of Udine and by the Istituto Sperimentale per la Zoologia Agraria, Ministry of Agriculture and Forestry, showed that four fumigations with Folbex VA in the absence of brood resulted in a 90-99 % mortality of Varroa mites (average value: 95.4 %). For our purposes an average value of effectiveness was fixed at 95 %. Taking this value into account, the number of survivors to 4 Folbex VA treatments carried out in absence of brood was calculated for each monthly group of hives. These numbers were then related to the numbers of mites found in the same colonies in October. The apiary treated in June was discarded because of the very low infestation.

Three other groups of 10 hives each were considered: two controls (A and B, the latter with a higher infestation than the former) and a set of colonies in which drone brood removal had been carried out. The rearing of the drone brood was stimulated by cutting the lower half of a central comb (1). In each hive, the drone brood was repeatedly removed (4-6 times) when most cells had been sealed.

In all colonies, inserts to recover dead mites were present on the hive floors from March 15 to October 1. Inserts were replaced every 15 days. In the control group A, the paper sheets were sometimes (cf. Fig. 3) ringed with vaseline.

To avoid bias in the estimate of the mean mite mortality, absolute values collected in a hive during the active season were put in a data vector on which the following algorithm was applied:

$$R = (A - L)/(H - L)$$

where:

R= uniformed value of mortality (Rmin=0; Rmax=1);
A= absolute value of mortality relative to a certain 15 day period;
L, H= lowest and highest value of mortality observed during the active season in the hive.

R values recorded at the same period in the hives forming a distinct set were then summed up together; finally, totals were uniformed again.

In the control group B, the numbers of honeybee adults was estimated every month according to the method described by Marchetti (8). Samples of adult bees, drone and worker brood were also repeatedly drawn out of these hives. Adult bees were forced into a plastic bag using a brush and a large necked funnel. About 250 bees were collected from a comb containing unsealed worker brood, other 250 from a comb filled only with honey. Samples of adult bees were examined by flotation (98 % ethanol for 30 min.). Samples of worker brood (about 170 cells per hive and period) were always removed from a central comb at the stage of pupa with brown eyes. To stimulate the rearing of a drone brood, the four corners of a central comb were cut away in each hive. The triangular spaces so created had sides 8-10 cm long. The cut comb was reintroduced into the nest in the same position. If present, sealed drone brood was removed 15 days after the date of cutting. Worker and drone brood cells were visually examined one by one.

In order to determine the rate of the natural mite mortality throughout the active season, the number of mites recovered from inserts in the 15 day period preceding the beginning of the operations in the monthly sets was divided by the total of mites that at the time of extraction of the inserts could be considered as already programmed inside colonies; this total was computed as follows:

$$T = AB + DB + F1 + F2 + (F3 + F4 + F5 + F6)/0.95$$

T = total number of mites;
AB = mites found in the sample of adult bees;
DB = mites found in the sample of capped drone brood;
F1, F2 = mites captured with the 1st and 2nd treatment with Folbex VA;
F3, F4, F5, F6 = mites captured with the 1st, 2nd, 3rd, and 4th treatment
                with Folbex VA in absence of brood.

Then, to the rate observed in the period March 15 - April 1 a value equal to 1 was given.

All the colonies were located in the province of Udine (north-east Italy) under the same climatic conditions; external food sources were abundant up to the end of July; no shortage of honey or pollen was however noticed inside the colonies during the following months.

For legal (if not ethical) reasons, the monthly sets of colonies were not moved after treatment towards locations where varroatosis was yet to be established; consequentely, some parasites probably entered in our hives from the surrounding apiaries. All groups of colonies were located in villages where the disease had appeared the year before, with the only exception of the set treated in August which was located in an area where varroatosis was present since two years. To fairly judge the possible extent of reinvasion in this apiary, it must be remembered that in September drones had already disappeared and flight activity of bees was low.

In the Friulian Region, during the spring and summer of 1984, the amount of rainfall was higher than the average (953 mm vs. 883 mm)(6); the number of rainy days was 69 out of 186. At mid-September outdoor temperatures decreased rapidly so that no brood was reared after this period.

3. Results and Discussion

The trend of the ratio between the numbers of mites found in October and that of survivors to four Folbex VA treatments, carried out in different months on brood-free honeybee colonies, is shown in Fig. 1. The analysis of regression indicated the exponential model as the best model

148

for good data fitting; the relative F value was highly significant (P=0.001). As a first impression, it would seem that deviations from linearity are due to the absence of factors limiting the mite population growth but also to the presence of higher rates of reproduction in spring. Anyway, there exists some evidence suggesting that drone brood is important but probably not essential for a rapid growth of the mite population to occur. In Fig. 2 the numbers of mites collected every 15 days from inserts in three groups of colonies are reported. In the set of hives where the drone brood removal was carried out, the increase in the mite population was notably delayed. Although in spring these colonies were more infested than those belonging to the control group A, in October the observed number of mites was lower. In fact, for this group the average number of mites recovered following treatment in a colony was 88.6 vs. 103.4 and 358.1 for control groups A and B, respectively. However, the infestation in the colonies submitted to biotechnical control was not reduced enough to make this practice advisable as the only control measure; this conclusion is in agreement with those of Abakumov (1), Mel'nik and Muravskaya (9), Santas and Lazarikis (18). The delay in the increase of the mite population was confirmed when data were transformed according to the algorithm given in Materials and Methods (Fig. 3).

The extent of reinfestation might be influenced also by a varying rate of the natural mite mortality throughout the year. Actually, in all the diagrams showing the trend in the number of mites recovered from inserts during the active season (Fig. 3), two facts can be noted: i) a peak in mortality in early spring, already observed in a former experiment (14), and ii) a decrease in mortality from mid-August. Both figures occurred with a frequency significantly (P=0.05) higher than the expected and were confirmed by the study on the rates of the natural mite mortality during the active season (Tab. I). When a value equal to 1 was given to the rate of mortality observed in the period March 15 - April 1, then the rate found at the beginning of autumn was just 0.23. It should be emphasized that in the second half of August the natural mortality rate was already decreased compared to those recorded in June and July (Tab. I). It is quite interesting to note that this decrease was related to important changes in the entity of other parameters. We refer to: i) the amount of drone brood which in August fell to zero (in Fig. 4 the amount of drone brood is indicated by the number of rebuilt corners in which a capped drone brood was present); ii) the number of mites per thousand cells of sealed worker brood (Fig. 4), and iii) the ratio between the number of mites found on 250 adult bees recovered from a central comb containing unsealed worker brood and that found on the same amount of adults collected from a (lateral) comb filled with honey (Fig. 5).

The trend of infestation observed for worker brood was similar to that recorded in Poland by Romaniuk and Duk (15). In both cases, when adult bees ceased to rear drone brood, there was a remarkable increase in the number of mites found inside worker brood. During spring, this number was low not only because of the high preference ratio of drone brood over worker brood (19) but also because of the lower size of the mite population; it should be remembered that during spring worker brood is

149

much less infested than drone brood but the former is extremely more represented inside honeybee colonies.

It appears from the graphs (Figs 4 and 5) that the number of mites on adult bees and inside the brood was largely compatible with a normal development of the colonies; hence, the above-mentioned results were not determined by a mite overconcentration, which is often observed with higher infestation levels during late summer, when the quantity of worker brood and adult bees is reducing. The reduction in the mortality rate and the change in the mite distribution should therefore be regarded as normal figures when autumn is approaching. Beekeepers should be informed of the decrease in the natural mortality rate so as to consider the presence of just 10 mites/insert in the two weeks preceding the overwintering of the colonies as an index of a possibly high infestation. Actually, when we found no mites on the last insert, chemical treatments revealed the presence of up to 321 mites; when the number of mites recovered from an insert was 1-5 and 6-12, we captured up to 427 and 1056 mites, respectively.

In regard to the feasibility of biennial instead of annual treatment against varroatosis, values reported in Tab. II seem to indicate that four fumigations with Folbex VA in the complete absence of brood are not fully sufficient, especially when the degree of infestation before treatment is high. On the other hand, the similarity between Varroa mite and honeybee with respect to the winter lifespan (21, 3), and the frequent occurrence of a brood rearing during winter would tend to exclude a significant reduction in the mite population size during the cold season. Therefore, the application of expensive or complicated and time-consuming methods to obtain brood-free colonies during the active season, in the view of queen replacement and chemical treatment, should be put into discussion, also considering the possible honey contamination with pesticide residues (2). In our opinion, only autumn treatments, carried out under conditions of decreasing brood, should be recommended for the northern and central part of Italy.

Tab. I - Relative rates of the natural mite mortality observed throughout the active season of 1984.

OBSERVATION PERIOD	RELATIVE RATE OF MITE MORTALITY
MARCH 15 - APRIL 1	1.00
APRIL 15 - MAY 1	1.08
JUNE 15 - JULY 1	0.62
JULY 15 - AUGUST 1	0.69
AUGUST 15 - SEPT. 1	0.48
SEPT. 15 - OCT. 1	0.23

Tab. II - Average numbers of mites /colony captured in the monthly sets of hives during the spring or summer treatment period and at the beginning of October.

PERIOD OF TREATMENT (END OF:)	MITES CAPTURED	MITES IN OCTOBER
APRIL	18	109
MAY	41	83
JULY	409	261
AUGUST	1193	231

REFERENCES

1. ABAKUMOV A. M. (1980). Building frames against varroatosis. Veterinariya, Moscow, no. 2, 40.

2. BARBINA TACCHEO M., DE PAOLI M., MARCHETTI S., D'AGARO M. (1986). Bromopropylate decay and residues in honey samples. In these Proceedings.

3. BREM S., KOPP H., MEYER P. (1983). The natural winter losses of Varroa jacobsoni in comparison with the whole mite population of the honeybee colony: an investigation of six colonies. Tierartzliche Umschau, 38 (1), 16-21.

4. CLAERR G. (1984). Prospects for a biological control method to control varroatosis. Atti del convegno internazionale dell'apicoltura, Lazise 1983, Studio Edizioni, 78-87.

5. ESPINOZA CAMARENA J., MEZA PECH I., LINERA D. R. (1981). Winter combined method of control of Varroa jacobsoni Oud. Proceedings of the XXVIIIth International Congress of Apiculture, Acapulco, Apimondia Publishing House, 329-332.

6. GENTILLI J. (1964). Il Friuli: i climi. C.C.I.A., Udine, pp. 595.

7. KOENIGER N., SCHULZ A. (1980). Experiments on a biological treatment of varroatosis by the control of all newly emerged bees. Apidologie, 11 (2), 105-112.

8. MARCHETTI S. (1985). Il "metodo dei sesti" per la valutazione numerica degli adulti in famiglie di Apis mellifera L. Apicoltura, 1, 41-61.

9. MEL'NIK V. N., MURAVSKAYA A. I. (1981). Drone brood combs and Varroa jacobsoni infestations. Veterinariya, Moscow, no. 4, 50-51.

10. PFEFFERLE K., D'ALLEMAGNE R. F. (1982). Le principe de la rotation dans la conduite des colonies et systeme d'obtention de colonies indemnes de Varroa. Apiacta, XVII, 64-70.

11. PETROV S. G., KHAZBIEVICH L. M. (1980). A biological trap as a method for controlling Varroa infestations. Doklady TSKhA, 266, 139-141.

12. RADCHENKO A. K. (1981). Experiences in the control of Varroa jacobsoni infestations. Veterinariya, Moscow, no. 4, 51-52.

13. REZINKIN G. S. (1978). Acaricides treatment of honeybees infested with Varroa jacobsoni. Veterinariya, Moscow, no. 3, 75-76.

14. RITTER W., RUTTNER F. (1980). Diagnoseverfahren. Allgemeine Deutsche Imkerzeitung, 14 (5), 134-138.

15. ROMANIUK K., DUK S. (1983). Seasonal dynamics of the development of Varroa jacobsoni in untreated honeybee colonies. Medycyna Weterynaryjna, 39 (12), 725-727.

16. RUTTNER F., KOENIGER N. (1980). A biological method for the elimination of varroatosis from honeybee colonies. Allgemeine Deutsche Imkerzeitung, 14 (1), 11-12.

17. RUTTNER F., KOENIGER N., RITTER W. (1980). Brutstop und Brutentnahme. Allgemeine Deutsche Imkerzeitung, 14 (5), 159-160.

18. SANTAS L., LAZARIKIS D. M. (1984). Using drone brood in the control of Varroa disease of bees in Greece. Entomologia Hellenica, 2 (2), 63-68.

19. SCHULZ A. E. (1984). Reproduction and population dynamics of the parasitic mite Varroa jacobsoni Oud. and its dependence on the brood cycle of its host Apis mellifera L. (Part I). Apidologie, 15 (4), 401-420.

20. SHILOV V. N. (1980). Effectiveness of biotechnical methods against varroatosis. Pchelovodstvo, 7, 19-21.

21. SMIRNOV V. M. (1978). Lifespan of winter generations of the Varroa mite. Pchelovodstvo, 12, 14-15.

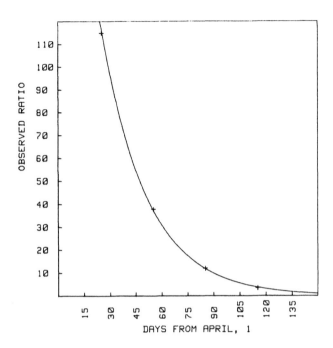

Fig. 1 - Ratio between the number of mites found in October and that of survivors to four treatments with Folbex VA carried out in absence of brood.

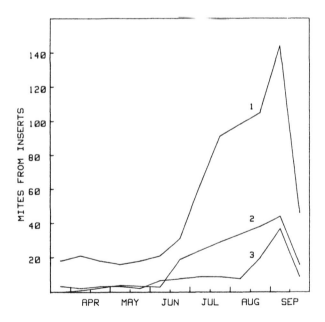

Fig. 2 - Number of mites collected every 15 days from inserts in the control group B (curve no. 1), A (curve no. 2), and in the set of colonies where drone brood removal was carried out (curve no. 3).

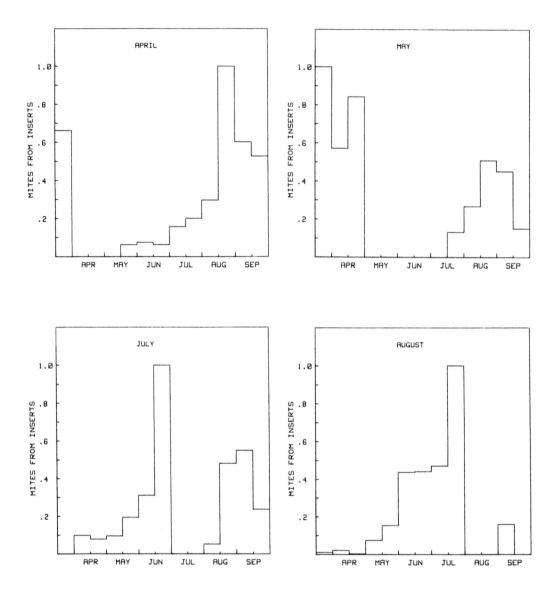

Fig. 3 - Uniformed number of mites collected every 15 days from inserts in
the monthly sets of hives, in the two control groups and in the
set of colonies where the drone drone removal was carried out
(arrows in the graph relative to control group A indicate the 15
day periods in which the inserts on the bottom board was ringed
with vaseline).

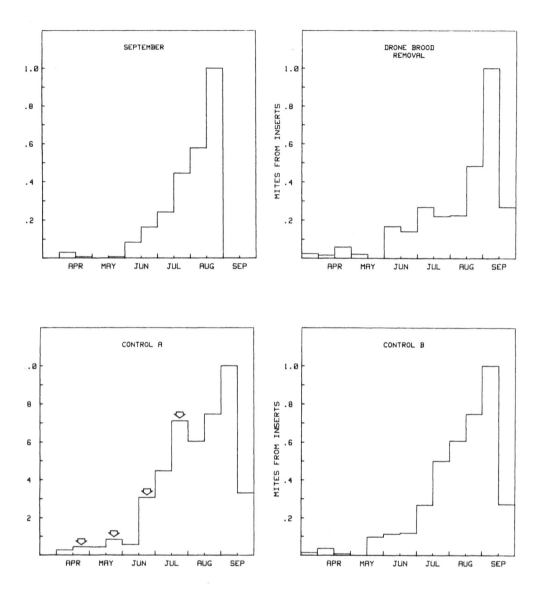

Fig. 3 - (continued).

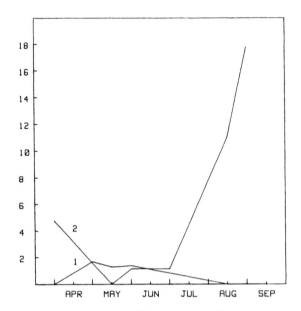

Fig. 4 - Average number of rebuilt corners of comb where a capped drone brood was present 15 days after the date of cut (curve no. 1), and number of mites per thousand cells of sealed worker brood (curve no. 2).

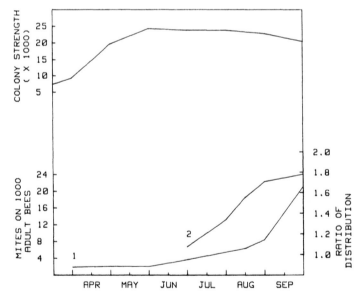

Fig. 5 - Average colony strength, number of mites on 1000 adult bees (curve no. 1), and ratio between the number of mites found on 250 bees recovered from a central comb containing unsealed worker brood and that found on the same amount of adults collected from a comb filled only with honey (curve no. 2).

156

Experiments with mite resistance to varroacidal substances in the laboratory

W.Ritter & H.Roth

Tierhygienisches Institut, Freiburg, FR Germany

Summary

The development of a mite resistance is of great importance for the fight against Varroatosis. A laboratory method was developed to determine the sensitivity of mites to varroacidal substances. Varroa mites became resistant in small beecolonies, when they have been treated with an underdose of Folbex VA and K79. In contrast no change of mite sensitivity has been found in colonies after 4 years of repetitive treatment with Folbex VA.

1. Introduction

The fight against Varroatosis, a parasite of the honey bee is almost only possible with the help of chemotherapeutic measures (3). Some preparations, which have been in use for several years, are not sufficiently successful anymore. This includes Phenothiazin, Tedion and Danikoroper (2). The mites have most likely developed a specific resistance. The development of a resistance must be recognized early in order to take the appropriate measures. The average sensitivity of a population is to be tested using K79 and Folbex VA and simple laboratory procedures.

2. Materials and Methods

Small colonies were made from relatively strongly infested colonies in order to examine the development of resistant mite strains. The small colonies consisted of 500 - 700 bees in Kirchhainer mating boxes and 2000 - 5000 bees in nucleus hives. K79 contains Chlorodimeformhydrochloride and Folbex VA Bromopropylate as active ingredients (5,6). 1/20th of a Folbex VA smoke strip was burnt, or 5ml of a K79 solution (0.5mg a.i.) was trickled onto the bees in mating boxes. The nucleus were treated with 1/8th of a smoke strip or with 40ml K79 solution (4mg a.i.). Treatment was repeated 3 times with 3 - 5 week intervals.

Varroa mites were collected off the bees 3 weeks after each treatment and the LC_{50} of each preparation was determined. The mites were dorsally fixed upon a slide. 0.2ul of each test solution was ventrally applied to the mites using a micro syringe (4). The test solution could be taken up percutaneously as well as orally by the mite.

The treated mites were kept in a climatic chamber at 16°C with 98% relative humidity. The mortality was determined after 24 and 48 hours. Mites were considered dead when they showed no reaction towards tactile stimulation of the first leg pair. In a series of experiments, three test solutions of different concentrations were simultaneously examined on 20 mites along with a control group treated with the same solution minus the active substance. The LC_{50} was calculated with a probit analysis (1).

In a further experiment, full colonies were treated in succeeding years with Folbex VA according to the manufacturer's instructions. Each treatment consisted of 4 applications of a smoke strip with 4 day intervals in broodless colonies. One colony each has been treated every fall since 1981, 82 and 83, respectively. A further colony, treated since 1981 was examined in the summer of 1985. The mites from the 4 colonies were examined in the same manner as those taken from small colonies.

3. Results and Discussion

A rapid development of mite resistance against K79 and Folbex VA could only be reached by creating optimal experimental conditions. Thus all colonies had larger numbers of unsealed and sealed brood during this time. The mites reproducing in brood cells only receives a relatively small amount of the medicament. The treatment was also carried out in a strongly under-dosed range.

4 to 5 weeks lay between individual treatments in order to ensure that the mites remaining on the bees were able to produce offspring. The number of mites reduced drastically with increasing frequency of treatment so that the removal of mite samples became difficult. There was, therefore, a relatively small range of samples.

The LC_{50} values determined in the lab for mites treated with K79 are presented in Table I and II. The mites originated from colonies treated at various frequencies.

Table I: LC_{50} of mites from colonies in mating boxes treated with K79

treatment	0.	1.	2.	3.
number of mites	60	90	57	62
LC_{50} 24h (ug As/mite)	1.50	6.02	8.13	7.24
LC_{50} 48h (ug As/mite)	0.50	1.77	2.24	2.72

Table II: LC_{50} of mites from colonies in nucleus hives treated with K79

treatment	0.	1.	2.	3.
number of mites	60	58	60	60
LC_{50} 24h	1.50	6.38	7.94	7.08
LC_{50} 48h	0.50	3.08	2.48	3.02

The values from both colonies correlated well with one another. The LC_{50} increased significantly in treated colonies when compared with untreated colonies. This was valid for values determined at 24 and 48 h. The more the treatments were repeated, the stronger was the tendency for an increase in LC_{50} values. The difference between the groups treated at various frequencies is smaller and not significant than their difference to the initial sample. The mites were already significantly more sensitive after the first treatment which may be evaluated as the beginnings of resistance. This became increasingly stable with further treatments.

A basically similar picture emerged from mites treated with Bromopropylate. In comparison to the initial sample the mites had a higher sensitivity from the colonies treated with Folbex VA and deriving from nucleus hives and mating boxes (see table III and IV). The difference to the initial value after the first treatment was significant, but not as clear as with K79. The sensitivity stabilized with the second treatment and increased significantly to more than 16-fold with the third treatment. This can also be taken as evidence for the development of resistance.

Table III: LC_{50} of mites from colonies in mating boxes treated with Folbex VA

treatment	0.	1.	2.	3.
number of mites	120	60	61	58
LC_{50} 24h	0.03	0.13	0.13	0.48
LC_{50} 48h	0.02	0.14	0.16	>0.48

Table IV: LC_{50} of mites from colonies in nucleus hives treated with Folbex VA

treatment	0.	1.	2.	3.
number of mites	120	59	90	60
LC_{50} 24h	0.03	0.13	0.13	>0.28
LC_{50} 48h	0.02	0.13	0.15	>0.28

Only the results recovered from the small colonies, which reflect possible development of resistant mites, are shown. The potency of individual preparations can be determined as a comparison.

The results here cannot simply be put into practice; this is shown by the values determined for Folbex VA upon full colonies. In this case, the interval between treatments was one year. Furthermore, broodless colonies were treated according to the instructions of the manufacturer, that is, 4 times with a smoke strip with 4 day intervals. Table V lists the LC_{50} values for Bromopropylate. The LC_{50} values increase in comparison to the initial sample. Thereby making the mites 4 - 5 times more sensitive after a treatment. In those colonies treated also 2 and 3 years ago, the sensitivity of the mites correlated to mites from untreated colonies. A new inspection, a year later, of colonies treated a fourth time support these results.

Table   V: LC_{50} of mites from full colonies treated with
            Folbex VA

            annual treatment
            since years)*   0   1   2   3   4
            number of mites   120 209 100 90   200
            LC_{50} 24h   0.03 0.12 0.03 0.03 0.03
            LC_{50} 48h   0.02 0.12 0.03 0.03 0.03

)* 4 applications of one Folbex VA strip each year

Under conditions used in practice no mite resistance
appears after 4 years of repetitive treatment with Folbex VA.
It becomes clear here what importance the roles of dosage and
of broodstate of the colony play. Frequent repetitive
treatments of colonies with brood rapidly increase the
development of resistant strains. This is most likely true for
other preparations, as well, which are used in fighting
Varroatosis.

REFERENCES

1. Cavalli-Sforza, L. (1974). Biometrie. Gustav Fischer
   Verlag, Stuttgart, 181
2. Rademacher, E. and Geisler, E. (1984). Die Varroatose der
   Honigbiene. Schelzhy and Jeep, Berlin, 104
3. Ritter, W. (1981). Varroadisease of honeybee Apis
   mellifera. Bee world 62(4),141-153
4. Ritter, W. (1986). Laboratory tests of chemotherapeutical
   substances in controlling varroa disease. Z.angew.Ent. im
   Druck
5. Ritter, W., Perschil, F. and Czarnecki, J.M. (1983).
   Treatment of bee colonies with isopropyl-4,4-
   dibrombenzilate against varroa disease and acarine disease.
   Zbl.Vet.Med.B. 30, 266-273
6. Ruttner, F. and Ritter, W. (1981). Eine Methode zur
   Varroatosebekämpfung über das Futter. Apidologie 12, 75-77

Diagnostic and therapeutic methods adopted in Sardinia against *Varroa jacobsoni* Oud.*

R.Prota
Istituto di Entomologia Agraria, Università degli Studi, Sassari, Italy

Summary

Fumigation with bromopropylate, examination of the sealed brood and adult bees inside and outside the hive were tested for diagnosis. The latter proved to be the most practical and economical, and preferable in the case of rapid investigations especially in rustic bee-keeping and wild populations. Therapeutic methods tested suitable to the typical climate of the island were bromopropylate fumigation with removal or blocking of brood at the end of summer, and use of natural substances in spring. Some satisfactory methods practised by bee-keepers are discussed.

1. Introduction

Many factors have favoured the spread of **Varroa jacobsoni** Oud. in Sardinia, particularly in the 70% prevalent cork skeps (rustic hives), which also supply swarms to the more modern apiaries. The progressive and rapid diffusion from the south has been due on the one hand to lack of preparation by the bee-keepers, and on the other hand to insufficient attention by the Health Authority in promoting adequate detection and control operations. In little more than 3 years from the first indication (9, 11), infestation by the mite has spread to almost every apiary, decreasing in intensity towards the higher altitudes (Fig. 1).

In coastal and southern regions, apiaries have suffered hive losses varying from 70% to 100% - and this in only a few years. It would seem that environmental contions are determinant, even where the bee-keeping is correctly carried out, in that they conduce to an almost constant brood presence throughout all the seasons, enabling the disease to accelerate its development correspondingly. Thus, the detection and control methods adopted in the Region have to meet the requirements of both the preponderant apicultural system and pedoclimatic characteristics of the diverse biotopes.

2.1. Diagnostic Methods

In consideration of previously made observations, we thought it advisable to compare current widely used methods (examination of residues after bromopropylate fumigation, of adults taken from within the hive,

* Supported by a grant from Ministero Pubblica Istruzione.

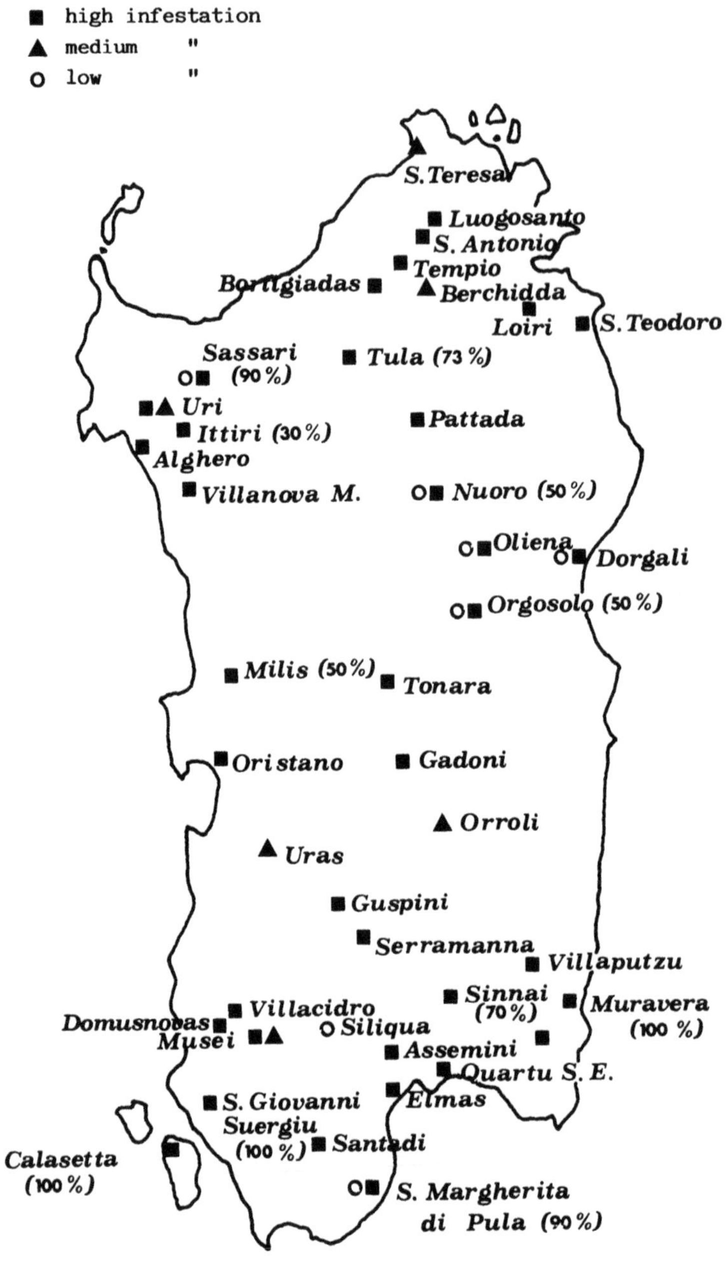

Fig. 1. Localities in Sardinia where **Varroa jacobsoni** Oud. has been noted
and estimated percentages of colonies lost.

and of the sealed brood) with another more practical, more economical method, above all more in keeping with the larger part of our apicultural patrimony based on the cork skep and wild honey-bee. The method in question consists of examining adults (mostly nectar-gatherers) taken from outside the hive entrance.

2.2. Description

Treatment with bromopropylate. - A strip of Folbex Va was suspended in a transparent cylinder placed over the food chamber hole, and the hive entrance plugged, being reopened an hour later. The vaselined sheet, laid on the hive bottom before fumigation, was taken out after 48 h. operational time averaged ca. 15 min per hive.

Sealed brood sampling. - Using a sharpened tube, about a 100 cells were taken, mainly female. Laboratory inspection included cell interior surfaces as well as larvae and pupae. Field work averaged 10 min per hive.

Sampling from within the hive. - Approximately 400 bees were sampled per hive by brushing them off the combs, and were treated to detach the mites. Sampling was slowed down by difficulty in isolating the queen; on average, it took about 20 min per hive.

Sampling from hive entrance. - About a 100 bees were brushed from around the hive entrance into a plastic bag. After treatment with ethylacetate they were freed from the mites by agitation in warm water (60°C). The method was rapid and practical, sampling time averaging under 5 min/hive.

2.3. Results and Discussion

All the methods adopted were efficient in clearly revealing that the parasite was present, even at low levels of infestation. Treatment with bromopropylate undoubtedly showed excellent detection efficiency, but was limited, according to our observations, in enabling the degree of infestation to be deduced (1, 6). Brood infestation varied considerably and seemed to bear little relation to that of the adult bees. The explanation lies in the way the mites distribute on the brood (10), discovered after an exhaustive examination of a destroyed colony. Statistical elaboration of the data obtained from the brood cells showed highly significant bunching in the distribution of the parasite (at least in medium-high levels of infestation), which from the first results would appear to fit a negative binomial model. Moreover, two degrees of bunching seemed to be indicated, namely at the cell and at the comb. Although these considerations need going into more deeply, they already sufficiently explain why the sealed brood method is unsuitable for deducing the degree of infestation (4). With regard to sampling from within the hive, the method gave particularly reliable indications as to the degree of infestation due to the size and homogeneity of the samples. Sampling from the hive entrance produced infestation percentage values greatly inferior to those obtained from within the hive, but in close relation to the latter (Tab. I). While an explanation of these lower values is mainly statistical and concerns the relatively few bees per sample, one cannot exclude the effect produced by the predominant presence of nectar-gathering worker-bees, longer-lived, more active, more robust and therefore presumably less infected. Nevertheless, this method is of the utmost practical and economic importance. Extended throughout our apicultural patrimony, it would readily enable infestation intensity and diffusion to be monitored in our Region. Statistical elaboration shows significant dependence, confirming the possibility of estimating adult infestation, even at low levels (<6%), by means of this external sampling method. Information would thus be swiftly and economically available for expedient and opportune therapeutic intervention.

Tab. I - Infestation on adult bees outside and inside the hive.

Infestation (%)

O/S hive	I/S hive
0 - 2	10.45 $\pm$ 6.4
2 - 6	14.20 $\pm$ 5.0
6 - 32.5	32.20 $\pm$ 14.5

Another point of interest revealed by the statistical analysis was the close dependence between adult and brood infestation (Tab. II) when the latter was measured by considering the number of individuals parasitized instead of the number of parasites.

Tab. II - Infestation (%) on (A) adults and (B) brood in 6 destroyed colonies.

Hive	A	B
1	42.56	42.00
2	37.50	46.00
3	50.00	52.80
4	32.50	33.60
5	53.98	49.50
6	29.30	30.75

The methods adopted did not present particular difficulties or require special equipment and can be applied by normally skilled beekeepers without aid from outside technical organizations.

3.1. Therapeutic trials

Basically, the control methods used did not differ from those widely known and practised. Application, however, had to take into account the pedoclimatic conditions in Sardinia and the consequently almost continual brood presence (Fig. 2). Three different methods were tested, mainly on the occasion of sanative work on heavily infested hives, namely:

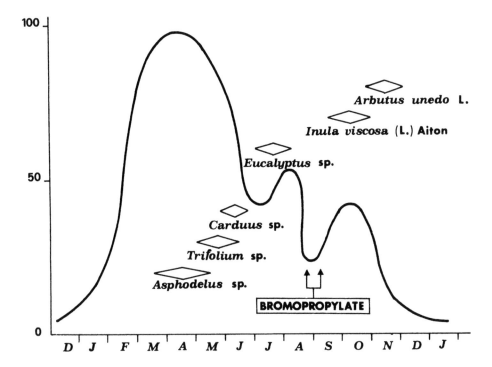

Fig. 2. Trend of brood egg presence during the year in northern Sardinian areas with apiculturally important flowering plants; (↑__↑) recommended period for chemotherapeutic intervention.

1. Treatment with bromopropylate (Folbex Va) after removal of brood;
2. Integrated intervention (brood interruption + chemotherapy);
3. Treatment with natural substances.

3.2. Trial 1

This trial took place in September, when practically the entire brood (at most 1 or 2 combs) could be removed. The remaining pupulation underwent bromoprylate treatment (1 strip per colony) in accordance with the normal procedure suggested by the manufacturer. The first treatment (also diagnostic) was followed by others according to the degree of infestation. The results obtained (Tab. III) show that the method was most effective, especially in the very heavily infested hives; but as shown in the table, the normally advised number of treatment did not suffice to bring down the attacks to minimum levels.

3.3. Trial 2

The trial began in September (12, 3) with temperatures around 20°C. In 5 hives, the queen was confined to the first frame on the left side by substituting the penultimate one with a vertical queen excluder pre-

Tab. III – Results of bromopropylate treatment (Folbex Va) in heavily
         infested hives (brood removed) – Sept 1985.

Hives	1	2	3	4	5
1st treatment	4035	3787	1200	518	1000
2nd　　"	1091	384	317	60	769
3rd　　"	385	142	39	30	119
4th　　"	109	173	78	23	320
5th　　"	42	36	13	8	104
6th　　"					
7th　　"					
8th　　"					

pared for the purpose. The trial took 24 days, from 27th Sept to 21st
Oct. At the moment of liberating the queen, the only brood comb was removed and sent to the laboratory for inspection (Tab. IV). The examination
of all stages of individuals present and their degree of infestation
led to the conclusion that it would be possible to avoid final elimination of the frame and thus suffer no economic loss.

Tab. IV – Results of brood frame examination in 5 hives undergoing chemo-
         terapy intregrated with brood interruption – Oct 1985.

Hive	EGGS		LARVAE		PUPAE		MITES	
	Side A	Side B	Side A	Side B	Side A	Side B	Side A	Side B
1	320	268	268	654	912	759	7	10
	tot. 558		tot. 922		tot.1671		tot. 17	
2	89	275	134	31	520	674	37	8
	tot. 364		tot. 165		tot.1194		tot. 45	
3	357	230	151	305	727	533	39	10
	tot. 587		tot. 456		tot.1260		tot. 49	
4	392	26	112	145	270	361	3	14
	tot. 418		tot. 257		tot. 631		tot. 17	
5	21	19	104	158	958	831	64	28
	tot. 40		tot. 262		tot.1789		tot. 92	

Egg laying restarted the same day in the central combs, but sufficient
time elapsed before cell capping began (9 days) to allow 3 treatments
to be done, which achieved maximum effect. As the partial results so
far obtained demonstrate (Tab. V), this method greatly reduces the infestation without seriously weakening the colony (2). It could be integrated with other complementary techinical and biological measures, such
as frequent requeening, removal of male brood, and ensuring proper
protein nutrition, or any of the measures aimed at preventing plundering

and whatever else might weaken the colony. Such integration would favour the use of less toxic substances in the chemotherapeutic phase.

Tab. V - Results of bromopropylate treatment in 5 hives subjected to brood interruption - Sept to Oct 1985.

Confinement of queen to lateral frame -27-09-85
Liberation of queen -21-10-85

Hive	6	7	8	9	10
1st diagnostic fumigation (27-09-85)	12	12	69	4	405
2nd therapeutic fumigation	97	106	251	73	532
3rd therapeutic fumigation	24	34	127	70	256
4th therapeutic fumigation	18	19	16	20	62

3.4. Trial 3

Four hives were treated in this experiment; a fifth, untreated, served as control. The mixture of natural substances used consisted of 10 g of camphor, 5 g of thymol, 5 g of menthol and 12 ml of vegetable oil. Dosage was 10, 20 and 30 ml for weak, medium and strong colonies respectively. The mixture was applied by soaking cotton-wool placed in a plastic container with small perforations in the lid. The treatment was carried out for a month during which vaselined frames were inserted every three days to collect the mites. The results reported in Fig. 3 show mite casualties on average to be greater than in the control. However, treating the hives with bromopropylate at the end of the trial produced unequivocally greater mite casualities than previously, thereby demonstrating the limitations of the mixture used (5).

4. Therapeutic operations in practice

In the case of colonies not heavily infested, measures can be taken (differentiated according to the consistency of the colony and therefore to their capacity for resisting parasitization) which for time and method of application (with the usual bromopropylate) would seem to satisfy both the health aspect and the bee-keeper's production requirements (7). Since the trials were conducted in a producing apiary and not under strictly experimental conditions, we laid our plans respecting a general principle of beekeeping practice; i.e. to reunite weak colonies and form nuclei for reconstituting the apiary's consistency. The consequent mixing of the populations with their different degrees of infestation allowed a general evaluation of the average degree of infestation throughout the apiary, but not a specific examination of single cases. Forming adequately disinfested nuclei not only serves to maintain the apiary strength, but also to control swarming and to satisfy the need to profit from late-spring and summer production. Autumnal reunion (using the best queens) takes advantage of arbutus flowering for the production of bitter honey.

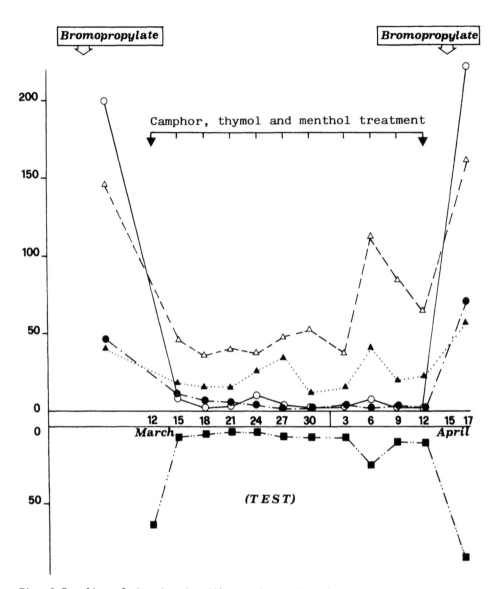

Fig. 3 Results of treatment with camphor, thymol and menthol, compared
with initial and final bromopropylate fumigation (4 colonies,
1985).

The operations, therefore, took place in summer at the moment
of minimum brood presence after nectar had been gathered from the
eucalyptus. Bromopropylate fumigation was used without the integration
of brood interruption or removal. After four fumigations, reuniting
the weaker colonies and post-winter formation of new nuclei began.
The trial could thus only make a general evaluation of the average
degree of infestation throughout the apiary, and not a specific examination of single cases. Diagnostic treatment followed the post-

168

winter formation of the nuclei, so that, if necessary, opportune corrective measures could be taken in time. The data in Tab. VI are expained as follows.

The chemotherapeutic treatment at the end of summer, normally used for the weaker colonies (occupying from 4 to 6 frames), produced an average fall of mites per hive of 768 at the beginning of the four fumigations and 161 at the end. The winter treatment, used for the stronger colonies (10 frames), produced a fall of 2250 and 271 mites respectively at the beginning and end of the four fumigations. The new colonies resulting from the reunion of the weaker colonies subjected to summer treatment were again fumigated in winter, and the mite average of 161 per hive was further reduced to 80. In 1986, the end of summer treatments (from the last days of August to the first days of September - in all, 12 days) with the usual bromopropylate resulted in a fall of 687 mites per hive at the beginning and 126 at the end. Compared with the previous year, colony strength had increased (now occupying from 6 to 8 frames instead of 4 to 6), and mite infestation had clearly decreased. Further testimony of the improved condition of the colonies lay in the absence of mortality, as occurred in August 1985. With regard to the choice of period for intervention, preliminary tests on 3 groups of 10 colonies per group made at progressively different times (27 Aug, 31 Aug and 16 Sept for group 1, 2 and 3 respectively) showed that the first (27 Aug) was the most suitable; before, that is, the considerable increase in brood related to the intense flowering of the **Inula** in the area.

5. Conclusions

In view of the gravity of the situation in Sardinia and the urgent need for a census of infestation conditions throughout our apicultural patrimony, the method recommended for swift and economical diagnosis was sampling the bees from outside the hive. The most suitable therapeutic method proved to be the use of bromopropylate fumigation at the time of minimum summer brood presence (end of August), i.e. after eucalyptus inflorescence and before reuniting the weaker colonies.

The Commission of the European Community at once recognized the problems raised by **Varroa** infestation; but we would strongly suggest that more research be urged and supported in Southern Europe, where apiculture has biological, technical and economic particularities different from the countries providing the greater part of the information regarding diagnostic and therapeutic methods.

Tab. VI – Summary of data from trials carried out in Sardinia (1985-1986).

	EXPERIMENTAL APIARY 1985-86				PRODUCTIVE APIARY		
TRIAL No.	I	II	III		weak colonies (4-6 frames)		strong colonies (10 frames)
			nat.sub	folbex	end. sum.	winter	
COLONIES	5	5	5		15		30
ADOPTED METHOD	brood remove	brood interruption	natural substances		colony reunion nucleus formation		colony reunion nucleus formation
ACARICIDE USED	bromopropylate	bromopropylate			bromopropylate		bromopropylate
INTERVENTION TIME	end summer	end summer	spring			winter	winter
TRIAL DURATION (days)	12	24	30		12	8	6
MITE FALL (Avge.) |S.D.|	beginning (2.100) end (140)	(100) (27)	(18) (19)	(105) (120)	(768) 825,31 (161) |213,06|	(80) |45,92| 227 |215,65|	(2250) 996,50 (271) |227,00|

END SUMMER 1986

		PRODUCTIVE APIARY	
		medium colonies (6-8 frames)	strong colonies (10 frames)
COLONIES	beginning (687)...............	28 (687) |688,90|	
MITE FALL (Avge.) S.D.	end (126)...............	(126) |158,09|	

170

REFERENCES

1. BARBATTINI L., MARCHETTI S., D'AGARO M., (1983). Risultati comparativi di diverse metodiche per la diagnosi della Varroasi. Atti Conv. Inter. Lazise, 1983: 71-77.

2. CLAERR G., (1983). Prospettive per i metodi di lotta biologica alla Varroasi. Atti Convegno Int. dell'apicoltura, Lazise 1983, 78-87.

3. CZARNECKI J.M. et al., (1983). Traitement d'automne de l'acariose. Résumé du XXIX Congrès Int. D'Apicolture, Apimondia, Budapest, p. 196.

4. GONCALVES L.S. et al., (1983). Comparaison entre les taux d'infestation des adultes et du couvain d'**Apis mellifera** par l'acarien parasite **Varroa jacobsoni** Oud.. Résumé du XXIX Congrès Int. d'Apicolture, Apimondia, Budapest, p. 109.

5. KAMBOUROV G. et al., (1983). Les Produits chimiques et vegetaux dans la lutte contre **Varroa jacobsoni**. Résumé du XXIX Congrès Int. d'Apiculture Apimondia, Budapest, p. 127.

6. MANNINO A., (1983). Diagnosi e terapia della Varroasi. L'Apicoltore Moderno, 6, 197-201.

7. PERSCHIL F., RITTER W., (1983). Les essaims artificiels dans la lutte contre la Varroase. Résumé du XXIX Congrès Int. d'Apiculture, Apimondia, Budapest, p. 160.

8. PFEFFERLE K., (1983). Obtention de junes colonies indennes de Varroa et renonvellement des populations. Résumé du XXIX Congrès Int. d'Apiculture, Apimondia, Budapest, p. 162.

9. PROTA R., (1983). Sulla presenza di **Varroa jacobsoni** Oud. in Sardegna. Studi Sass., Sez. III, Ann. Fac. Agr., Vol. XXX, pp. 255-264.

10 ROSENKRANZ P., (1984). Distribution of **Varroa** females within the honeybee brood nest and consequences for biological control. Experts meeting CCE "Research activities on Varroatosis in the European Countries", Thessaloniki 26-28 Sept 1984 (in print).

11 SULIMANOVIC D. (1983). Le sort des colonies attaqués par **Varroa jacobsoni** Oud. dand le Nord owest de la Yougoslavie. Résumé du XXIX Congrès Int. d'Apiculture, Apimondia, Budapest, p. 196.

12 TASHIRO K., (1985). New strategy for more efficacions control of **Varroa jacobsoni** with ZPK (Phenotiazine preparation). Ab. rep. XXX Int. Apicultural Congress, Nagoya, p. 75.

171

Chemical control with Coumaphos (Asuntol) and Amitraz against *Varroa jacobsoni* in Sicily

C.Gallo
Istituto di Zootecnica, Facoltà di Agraria, Università degli Studi, Palermo, Italy
P.Genduso
Istituto di Entomologia Agraria, Facoltà di Agraria, Università degli Studi, Palermo, Italy

SUMMARY

Authors refer on the first investigations and on the good results obtained in Sicily against Varroa jacobsoni using Coumaphos (Asuntol) and Amitraz (Bumetran).
Varroa jacobsoni population decreased of about 95% and hives infected at the second level could be saved with 2 or 3 treatments.
Gaschromatographic analysis and honey samples showed values of residues much lower than those established as ADI.

Varroa jacobsoni was described for the first time in Sicily by Gallo and Aiello in December 1983, but it is to suppose that it was present from at least three years; from then on it has been found in the whole Island, causing very severe problems.

Beekeepers are looking for a good product against Varroa jacobsoni, being easy and quick to apply and to handle, cheap and harmless for the operator and well tolerated by the bees, which have not to change their behaviour, and associating to a security time for man and bees a great persistence for mites.

Among numerous products against Varroa, a particular attention was given to Coumaphos (1) and Amitraz (2). Coumaphos develops a very low toxicity for bees and a long persistence for mites. As known (F. Ruttner, W. Ritter, W. Götz, 1980), it has a systemic action and is distributed to hives by means of trophallaxis.

Asuntol was orally administered in a honey-watery suspension: 1 gr of Asuntol was diluted in 200 ml of liquid and 25 ml (40 mg of Coumaphos) were given to every hive containing 10 populated combs; Asuntol was given once to combs without brood and three times to brood-combs every 7 days.

(1) soluble power - sold by Bayer as ASUNTOL for veterinary use, containing 30% of active ingredient.
(2) many products are sold in Italy only for agricultural use: we used BUMETRAN Shering at 21.6% of active ingredient.

During a previous screening, with the product we used, it was observed that higher concentrations, orally administered, did not be the death for bees, which had a great mortality if sprayed, even if at lower concentrations.

17 mg· of Amitraz were given to every hive as aerosol for 60 seconds at a temperature of 38-40°C.

All treatments were made on December 15, 1985 and on February 25, 1986 on a first and on a second group, respectively.

It is to remind that broods are always present in the coastal areas of Sicily. Assays have been carried out on 13 severly infested colonies of sicilian bees.

Before and after treatment a sample of about 250 bees was collected and shaked for 40' in water and detergent accordingly to the technique developed by Koeni.ger to know the ratio bees/Varroa.

Honey was collected at the beginning of June and separately extracted hive by hive.

A sample of 50 g. of honey was gaschromatographically examined using the technique of Flamini and Robin (1985) with a Varian 3700 with E.C.D.-column G.P. 4%-S.E. 30 60%, 2 m lenght x 2 mm ∅, carrier gas : nitrogen-temperatures: column 220°C, injector 250°C, detector 300°C.

Results are reported in the following table:

FAMILIES	RATIO BEE/VARROA ANTE	PRODUCT	RATIO BEE/VARROA POST	RESIDUES mg·/Kg
1	20.33	Asuntol	death	-
2	16.14	Asuntol	2.79	0.050
3	10.13	Asuntol	2.50	0.045
4	15.14	Asuntol	3.15	0.038
5	9.16	Asuntol	1.90	0.048
6	11.96	Asuntol	3.10	0.023
7	13.96	Amitraz	2.80	0.017
8	32.15	Amitraz	death	queen-bee with 5 Varroa
9	63.31	Amitraz	death	-
10	12.75	Amitraz	2.65	0.16
11	13.40	Amitraz	3.00	0.26
12	15.30	Amitraz	2.30	0.21
13	8.25	Amitraz	1.85	0.096
14	14.60	--	death	-
15	16.80	--	death	-
16	12.70	--	death	-

Bees were of the black A. mellifera sicula maior species.

It is to remark that Coumaphos and Amitraz greatly reduce Varroa population even if bee colonies develop an infection reaching the second stage (10 to 20% of acari). Six colonies, including the control three, died because the infestion exceeded the third stage.

All surviving colonies were stimulatingly feed and produced from 10 to 15 Kg of honey.

Gaschromatographic chemical analysis on honey samples showed values of residues much lower than those established as ADI.

The mean concentration values of residues of Asuntol ($x = 0.041$ mg/kg; s.d. 0.011), that the standard classification describes having a lower toxicity than Amitraz and which is very cheap and easy to use, are about 4 times lower than those of Amitraz ($x = 0.15$ mg/kg; s.d. $= 0.09$).

Acknowledgements:

The Authors thank Dr. Antonino Corrao for the gaschromatographic analysis carried out at the Istituto Zooprofilattico Sperimentale della Sicilia - Palermo, and Mrs. Patrizia Tommasi and Mr. A. Corsino for their assistance in the field.

Bibliography

1) Flamini C. et Robin S.(1985)- Me'thodes dosage des résidus d'AMITRAZ dans les mieles - Bull.Lab.Vet.,88, 47-57.

2) Ritter W.(1986) - Versuche zur Entwicklung und Prüfung von PERIZIN einen systemischen Medikament zur Bekämpfung der Varroatose der Honigbiene - Tierärztl.Umschau, 41, 105 - 112.

3) Ruttner F.,Ritter W.,Götz W.(1980) - A.D.I.Z.,14,160-166.

4) Santas L. (1985) - Notions preliminaires sur l'utilisation du produit ASUNTOL contre le varroase - Apiacta 20, 33-37.

Thermo-chemical control of *Varroa jacobsoni* with minimal application of Amitraz

J.M.Kulincevic & F.Tomazin
Beekeeping Combine 'Beograd', Beograd, Yugoslavia

Summary

To control Varroa jacobsoni beekeepers in Yugoslavia are widely using chemical substance amitraz in a form of domestic fumigant „Varamit". To keep the parasite at low level repeated treatments are necessary in autumn and in certain cases during other periods of the year, when there is no honey surplus in the hives.
Application of thermo-chemical treatment with experimental device „Varoamat", which was constructed in a form of a mobile and portable floorboard for Langstroth-Root hive with built-in heater, air pump and instruments, enables successful prevention of Varroa jacobsoni by single treatment with a very small dose of amitraz. Insertion of air and active substance between frames with bees (which is a smoke obtained by burning „Varamit") is done by air pump. During the experimental work with „Varoamat" device, a research of bees' behaviour was done at the temperature of 42°C and lowered air humidity of 30 to 40%.

1.1 Introduction

Application of chemical preparation for prevention of Varroa jacobsoni specially when frequent and applied in larger dosis can bring a danger of residual contamination of honey, pollen and wax. Anyhow, the fact that beekeepers as well as other farmers mainly rely on chemical means because their application is far easier than it is the case with other methods, among which is thermal treatment of Varroatosis what actually means use of increased temperature of 46-48°C and lowered humidity up to under 30% of relative humidity with specially complicated and expensive procedure.

Thermal method for prevention of Varroatosis is according to Solovjeva initially introduced in USSR by Hrust in 1978. Unlike other countries endangered by mite, in USSR this way of Varroatosis prevention is widely used (Smirnov, 1985). In Soviet literature we find sufficient data dealing with advantages and disadvantages of thermal treatment of bee colonies (Komissar and Ponomar, 1982; Komissar and Pilecka, 1983; Komissar 1983; Burtov 1982).

Each method described and applied by Russians requires shaking of bees from the frames into a special screened cassets at the time when there is no brood in the hive and putting cassets into special chamber containing devices for heating and ventilation (Komissar, 1983). It must be strictly followed that the temperature does not raise over 46-48°C. Advantage of this type of mite prevention is that there is no contamination of honey, wax and propolis, but it requires high use of working power, i.e. three workers for 30 minutes length of time per one hive, while for application of phenothiazin tablet only 26 seconds is necessary and for spreading with Sineakar 50 seconds (Solovjeva, 1983.). Besides, for treatment of shaken bees with heat of 46-48°C rather variable results were obtained with efficiency between 50.2 to 76.0 percent. Also, the bees could be damaged by heat, while queen-bee stops laying eggs for few days.

Due to this variance in USSR combined thermal methods with use of some acaricides were applied (Smirnov, 1985), but also in that case bees had to be shaken out of hives. This way efficiency of 98% was reached by Varoatin, 96% with Phenothia-sin and 94% with Folbex VA.

To avoid high use of manpower for shaking bees we tried some preliminary investigations on combined use of heat and amitraz (Varamit) directly on bee colonies together with hive bodies and frames, all being subjected to the treatment.

1.2 Material and methods

In this experiment we used twenty bee colonies placed into Langstroth-Root hives with two bodies. Colonies were not equalized and sealed brood was removed from them beforehand. Queen-bees remained in hives.

To establish percentage of infestation treatment was done on 250 bees in separate cages (Kulincevic et.al. 1973) which after being filled with bees were put into the empty hive body above the Varoamat. Double dose of „Varamit" was used for 10 minutes at the temperature of 42°C. After 8 hours there were no parasites on bees.

Twenty experimental colonies were devided into four groups with five bee colonies in each group:

I.   Varoamat   - treated colonies
II. Varamit   - treated colonies
III. Fluvalinate   - treated colonies
IV. Control group

During experiment outside temperature was 18°C, relative air humidity 80%, sky clear and no wind.

Device for thermo-chemical treatment called Varoamat which was constructed by the second author of this report consists of low wooden box which during the application replaces the floor-board. An electromotor with 150 W, heater of 1500 W and air pump Ø 90 mm, were built into this box. On the front side there is a board with switches and behind them there is a thermometer and hygrometer. Electric power is brought through reley and switches-off via thermostat. Two wire nets are placed over the box. Upper one through which affected mites fall through and lower, thicker one on which parasites remain.

Bee nest was in one or two chambers during the treatment. Such chamber or chambers with bees and frames were put onto Varoamat and instead of covering board one shallow wooden box was used which contained the top thermometer. Part in which the thermometer was put was separated from bees by wire net.

After switching-on the electrical power in 3-5 minutes heating body gets warm and after that in the next 3 minutes by means of air pump warm air is blown among bees until on the top thermometer temperature reaches 40°C. At that point air pump is switched-off and Varamit burnt on the bottom thick wire net. In the following five minutes heater continues to work but air pump is switched-off. After that chambers with bees are put back into the place and quickly covered because effect of Varamit still continues. To protect respiratory organs during work mask of gauze and cotton was used.

Counting of dropped parasites is done on Varoamat net after 10 minutes and under hive on inserted pad after 24 hours. Also in this experiment plastic strips were used, with 10% Fluvalinate, length of 50 cms., and placed diagonally on wire net above the hive bottomboard.

For Varamit, Fluvalinate and Control group counting of dead mites was done after 24 hours.

1.3 Results

Results of this preliminary and in volume restricted experiment with combined thermo-chemical treatment of bee colonies infested with **Varroa jacobsoni** were shown on Table 1.

As it can be seen from Table 1. bee colonies were of various strength with around 8.000 to 17.000 working bees. Infestation of adult bees ranged from 1.38 to 2.79 percent. Single treatment with Varamit (Kulincevic, 1985) which was applied 24 hrs earlier removed in average 75.3 percent of parasite, what is in concordance with our experience when dealing with treated colonies without brood. Through application of Varamit by means of Varoamat an efficiency of additional 22.2 percents was reached, which all leads to an average efficiency of thermo-chemical treatment of around 97.5 percent. There is also variation here, which goes from 95.4 to 98.3 percent.

From Table no.2 we learn that plastic tape 50 cms long, containing 10 percent of Fluvalinate, did not have an effect on mite after 24 hrs. Number of dead mites on floorboard under the net was same as in control group. However, five days later there was mortality increase of mites in the colonies that were treated with Fluvalinate strips.

1.4 Discussion

As it is known that acaracides are more efficient at higher temperature and lower relative humidity (Komissar and Pilecka, 1983; Komissar, 1983) it could be expected that

TABLE 1. Treatment efficiency of bee colonies infested with
mite, by application of thermo-chemical method and
use of „Varamit" strips

Col. No.	Approx. No. of Bees	Appr. No.of Mites	% of infes.	„Varamit" treatm. dead mites	In %	„Varoa. tret. dead m.	In %	Efficienc of heat.-chem.tr.
256	17.200	231	1.38	184	77.6	42	17.7	95.4
291	15.000	214	1.42	157	73.4	50	23.4	96.7
273	12.700	355	2.79	263	74.1	86	24.2	98.3
262	8.000	207	2.62	154	74.3	48	23.2	97.5
286	9.400	153	1.92	116	75.8	32	20.9	95.4

TABLE 2. Comparative treatment of bee colonies infested by
mite with„Varamit", Fluvalinate and Control group

Number of dead mites after 24 hours					
Colony Number	Varamit	Colony Number	Fluvalinate	Colony Number	Control group
303	172	338	11	317	6
336	180	322	12	321	5
328	131	295	7	253	21
307	256	293	4	260	19
275	214	339	19	298	3

efficiency of Varamit will increase by use of Varoamat device.
It should be taken into account that according to our and
Turkish investigation, by three treatments of infested bee
colonies without brood, efficiency of Varamit amounts to
around 93 percents with possibility that during winter season
by additional treatments that percentage somewhat increases.
By use of Varoamat with single treatment multiple application
of amitraz can be avoided. This is specially important for
beekeepers who seldom change honeycomb in the hive, as the

treatment of brood nest with chemical substances against mite is more frequent, and in some countries where there is no brood rearing interruption, as it is the case in Israel (Efrat et al. 1986) where it is done by amitraz 45 to 50 times during one year, what is really too much.

Our experiment with Varoamat has shown that at the time when there is no brood in hive it is possible to reach satisfactory reduction of mite by single treatment only. True, this way also takes much more time unlike Varamit strips application.

Besides, some variations in results are noticable, what certainly depends on bee colony strength, presence of holes and cracks on hive and other conditions.

This device for thermo-chemical treatment can be successfully applied at shaken bees as it was done by Smirnov (1985) in USSR.

In package swarm production in U.S.A., when presence of mite is confirmed, which is only a question of time, one of the possible solutions would be the use of thermo-chemical method with amitraz for mass removal of parasites from adult bees.

Although there are very possitive reports about use and efficiency of Fluvalinate for mite prevention, Borneck (1986), our first trial with plastic strips did not give results 24 hours after treatment. It is possible that this preparation was meant for a long term treatment because five days later we had an increase in mite mortality.

References

1. Borneck, R., 1986 - Fluvalinate an interesting molecule in the battle against Varroa mites. Apimondia Inter. Symposium. Health protection of honey bees, Zagreb 1986, Yugoslavia.

2. Burtov, B.J., 1982 - Heat treatment and degree of infestation with mite. Pchelovodstvo 6, 15 (In Russian).

3. Efrat, C.; Jacobson, B.A.; Rosental, K. - 1986 - Varroa Control in commercial apiaries in Israel. Apimondia Inter. Symposium. Health protection of honey-bees, Zagreb, 1986.

4. Komissar, A.D.; Ponomar, A.A. - 1982 - Efficiency of thermal bee treatment. Pchelovodstvo, 8: 19-20 (In Russian)

5. Komissar, A.D., 1983 - Humidity regime at thermal bee treatment. Pchelovodstvo, 7 :15-16 (In Russian).

6. Komissar,A.D.;
   Filecka,I.V. - 1983 - Influence of thermal treatment on
                     Varroa jacobsoni mite.
                     Pchelovodstvo,5 : 17-18 (In Russian)

7. Kulincevic,J.M.;
   Rithenbuhler,W.C. 1973 - Laboratory and field measurements
                     of hoarding behaviour in honey-bee.
                     J.apic.Res.12: 179-182

8. Kulincevic,J.M.-1985 - Il „Varamit" controlla con molto
                     successo la varroasi in Yugoslavia
                     L'apicoltore moderno 76; 155-160
                     (In Italian - English summary).

9. Smirnov,A.M., 1985 - Combined way of thermal treatment.
                     Pchelovodstvo, 5 : 14-16 (In Russian)

10. Solovjeva,L.F.,1983 - Thermal treatment of bees infested
                     with Varroatosis.
                     Pchelovodstvo, 1 : 17-18 (In Russian)

Comparative studies on the effectiveness of Folbex VA[R], Amitraz and Formic acid in the control of Varroatosis in Poland

Z.Gliński & M.Chmielewski
Bee Diseases Research Laboratory, Agricultural University of Lublin, Lublin, Poland

A number of chemicals are available to control of Varroa jacobsoni but at present there is no commonly accepted and satisfactory chemical means of eliminating Varroa from honey bee colonies. Chemical treatment reduces only the number of mites on bees and does not destroy the mites on capped brood.

During 1984-1986 a systematic research was done on the efficacy of Folbex VA/fumigant strips, amitraz/0.0012% aqueous solution 250 ml per colony, and formic acid/98% evaporation in 120 colonies (40 colonies in each group), naturally infected by V. jacobsoni. In a half of colonies treated building frames for the drone brood were used twice in summer. Two treatments with Folbex VA or amitraz at 7 day intervals, and one treatment with formic acid, were applied in the spring and in the autumn in the absence or nearly absence of brood.

Folbex VA showed to posses an outstanding capacity for control of V. jacobsoni, and the infestation was reduced to 95-98% after two treatments in colonies in which a biological method of Varroa control was used. Only a little less effective in the control of V. jacobsoni appeared to be amitraz and formic acid used in conjunction, with the removal of the capped drone brood on building frames in summer.

Further work is in progress to examine new formulations and different methods of application of amitraz and formic acid.

Treatment of Varroa-disease with Fumilat in the East region of Poland

M.Chmielewski & Z.Gliński
Bee Diseases Research Laboratory, Agricultural University of Lublin, Lublin, Poland

The experiments were performed on 35 honey-bee colonies in the East region of Poland where the intensity of <u>Varroa jacobsoni</u> invasion is high. Like to Folbex VA[R] Fumilat fumigant strips containing brompropylat are developed in Poland for the control of <u>V. jacobsoni</u> mite.

In contrast to Folbex VA used four times at 4 days intervals in spring and autumn, Fumilat fumigant strips were used only twice in spring and autumn. Fumilat, especially when used in colonies nearly almost devoided of brood, revealed the same efficacy as Folbex VA. After two applications of Fumilat fumigant strips at 4 days intervals, a minimum of 94-95% of the mites present on bees were destroyed and no significant loss of bees was noted.

Present status of the control of Varroa-disease in Greece

L.A.Santas
University of Agricultural Sciences, Athens, Greece

Summary

Malathion as powder is considered and used in Greece as bee medicine. It is the main chemical for the control of Varroa disease in our country for many years and it is used with great success, in concentration below 1‰, applied mainly by hand between frames or over combs.

According to our research, Malathion has not shown any direct or side adverse effect to the honey bees or brood while no residues of it was traced in honey. In addition there is no evidence that mite developed any resistance to Malathion, although it is used for many years (since 1979).

1. Introduction

There are no accurate data in our country concerning the economic losses from Varroa disease in the bee-colonies. However during the first three to four years folowing the discovery (April 1978) of that serious disease, and mainly during the period 1981-82 the bee-colonies had been heavily damaged and many apiaries in various places were destroyed by this mite (4).

Later on and after the use of some acaricides the problem has restricted and now it is not so severe.

Malathion as powder is considered and used in Greece as bee medicine. It is the main chemical for the control of Varroa disease in our country for many years and it is used with great success as powder, in concentration below 1‰, applied mainly by hand between frames or over the combs (1,2,4).

In Greece and mainly in the Central and South parts, there is a severe problem in the control of Varroa disease. This is due to the fact that brood exists in the beehives all year round. However, during the winter the brood is restricted, at least for a short period, and sometimes there is no brood at all.

The control with Malathion in dust 1‰ a.i. at the quantity of 2 grs per hive or 0,5‰ a.i. at the quantity 5 grs per hive, in the winter and for 4-5 times at intervals of 2 days give very good results (1,4).

Using this method of Malathion it is possible to reduce the infestation by 98% - 100%, thus the infestation is so far reduced to such a level that no damage from the mites is observed until next winter (4).

According to research work in Greece (3,4,5) and Turkey (6) it is confirmed that Malathion does not contaminate the honey and does not leave dangerous residues in the wax. On the other hand the existing honeys in the hives during the period of Malathion treatments, that is in the winter, they are almost all consumed by the bees, besides a long interval exists between treatments and first harvest of honey.

The Malathion, as aforementioned, is characterized as bee medicine when it is used in the form of powder, with talc or sugar and in a concentration lower than 1‰. Thus the Malathion has widely been used to control Varroa disease since years in Greece, and it is the fundamental agent used by Greek apiculturists against this pest.

After such a long time of using of Malathion, it was too much interested to investigate and check the behavior of Malathion on bees (any side effects on bees) as well as to investigate any resistance development of the mite.

2. Materials and Methods

The research work was started in July 1985. Treatments were carried out with Malathion in beehives in order to compere the results of this test with the results of previous ones (4). For this reason, the data of the Table I of our work (4) was used, (mainly the first and second treatment). Also observations and samples were taken from beehives is mountenous areas. In these beehives the control of Varroa disease was done only with Malathion since years.

For this test ten one-chamber bee-colonies were used, located in the apiary of the Univ.of Agricultural Sciences of Athens (Votanicos-Athens).

The beehives were divided into two groups, consisted of five beehives each. Each of them received 1‰ a.i. and 0,5‰ a.i. Malathion respectively.

Five treatments were given to each hive on 7/12,9/12,11/12,13/12 and 15/12/85, in that period there was not brood in the beehives, so the treatment was made during a broodless period.

Before treatment, a special mite-collecting floor, was inserted into each hive. These special mite-collectings were checked for fallen mites 2 times between two applications.

Five days after the final application 200-300 bees were taken of each colony. The sampled bees washed out in a gasoline in order to count the number of mites remaining of them.

3. Conclusions

Considering the experimental data obtained in winter 1985-86 in connection with the results of the pertinent experimental work carried out in our country in the past (winter 1981-82) (4) and also in connection with the results of observations and samples, we can conclud that:
I. Beehives on mountainous areas (with long broodless period during the winter), on which Varroa jacobsoni is controlled only with Malathion powder (since 1981) were found free of mites or in very low infestation.

From 18 samples of bees and brood which collected from 18 different beehives,(which belonged in different apiaries),8 of them were found to be free of mites, while in the rest 10, a very low infestation was found in sampled bees or in the brood (Table I).

Treatments on those 8 beehives with Malathion, soon after the samples were taken, confirmed the aforementioned data, as no any mite was fallen.
2. In addition after observations on apiaries in the same areas and information from skilled apiarists the bee-colonies have not showed any abnormalities.
3. Compering the results of the winter of 1985-86 and those of 1981-82 one might be said that there is no evidence that mites developed any resistance to the Malathion although it is used for many years (since 1979) (Tables , II,III,IV,V).

REFERENCES

1. EMMANOUEL,N.G., SANTAS,L.A., TABOURATZIS,D.G., 1982. Varroa disease and its control in Greece. VI Intern.Congr.of Acarology, Edinburgh 5-11 Sept. 1982, pp. 1099 - 1105.

2. IFANTIDIS,M.D., 1980. Malathion als Kontaktmittel zur Bekäpfung der Varroamilbe pp. 144-149 from Diagnosis und Therapie der Varroatose. Apimond. Publis.House, Bucharest, Romania.

3. LIAKOS,B., 1983. Research of toxic residues of Malathion in honey, Veterin. Greek, 1:308-314 (in Greek).

4. SANTAS,L.A., 1983. Varroa disease in Greece and its control with Malathion. In proc. of Meeting of experts group "Varroa jacobsoni Ouds". affecting honey bees: Present status and needs" Wageningen, 7-9 February 1983, pp. 73-76.

5. THRASYVOULOU,A., IFANTIDIS,M., PAPPAS,N., 1983. Malathion residues in Greek honeys. 2nd Greek Apicult.Congr.Athens 15-17 Nov. 1983 pp. 144-149 (in Greek).

6. TUTKUN,E., UNAL,G., ILIKLER,I., INCI,A., YILMAZ,B., 1984. An experiment on the effectiveness of certain chemicals against Varroa jacobsoni Ouds in Central Anatolis, Turkey. 3rd Intern. on "Varroa" of Apimondia. Split (Yugoslavia), Sept. 30-Okt. 2, 1984.

Percentage of infestation in the bees and
in the sealed worker brood in samples which
were taken from beecolonies in mountainous
areas, where Varroa disease is controled
with Malathion.

Date	Area	Mites Bees	Infestation %	Infestation in worker brood(*) %
20/7/85	Karpenisi	1/280	0,3	0,0
	"	0/260	0,0	0,0
21/7/85	Pavliani	1/302	0,3	0,0
	"	1/289	0,3	0,0
9/8/85	Karpenisi	0/265	0,0	0,0
	"	0/285	0,0	0,0
10/8/85	Pavliani	0/293	0,0	0,0
	"	0/253	0,0	0,5
10/8/85	Gravia	1/321	0,3	0,0
	"	2/280	0,7	1,0
13/9/85	Karpenisi	1/285	0,3	0,5
	"	0/247	0,0	0,0
14/9/85	Agrapha	0/267	0,0	0,0
	"	0/279	0,0	0,0
28/9/85	Karpenisi	0/327	0,0	0,0
	"	1/307	0,3	0,5
29/9/85	Grevena	1/263	0,3	0,5
	"	0/280	0,0	1,0

* 200 worker cells checked

TABLE II

Number of mites fallen after application with Malathion 1‰ (2 grs) and Malathion 0,5‰ (5 grs) in 10 beehives at Votanicos – Athens Greece, during the winter 1985 – 1986.

Date	Treatment									
	Malathion 1‰					Malathion 0,5‰				
	1	2	3	4	5	1	2	3	4	5
8.12.85	1412	312	144	263	47	1512	325	201	84	42
*9.12.85	61	9	21	14	3	62	17	21	7	4
10.12.85	201	65	31	83	7	212	52	47	11	7
*11.12.85	24	8	5	17	2	41	12	7	4	2
12.12.85	41	24	3	9	1	55	17	3	2	1
*13.12.85	9	3	1	1	0	31	8	1	0	1
14.12.85	12	9	3	2	0	1	3	2	1	0
*15.12.85	1	0	0	2	0	0	1	0	0	0
16.12.85	1	1	0	0	0	1	0	1	0	0
17.12.85	0	0	0	0	0	0	0	0	0	0
T O T A L	1762	431	208	391	60	1915	435	270	109	56

Five treatment were given to each hive on 7/12, 9/12, 11/12, 13/12 and 15/12/1985.

T A B L E   III

Percentage of mites fallen after application with Malathion 1‰ (2grs) and Malathion 0,5‰ (5grs) in 10 beehives, at Votanicos - Athens Greece, during the winter 1981-1982 and the winter 1985-1986.

Winter	treatment	Malathion 1‰					Malathion 0,5‰				
		1	2	3	4	5	1	2	3	4	5
1981-82	I*	83,24	72,75	76,76	86,27	69,52	83,20	75,23	77,39	89,39	82,78
	II**	80,07	57,73	89,13	85,71	86,20	73,73	64,15	86,46	71,42	81,81
1985-86	I*	83,59	74,47	79,32	70,84	83,33	82,19	78,62	78,51	83,48	82,14
	II**	77,85	66,36	83,72	87,71	81,81	74,19	68,81	88,52	83,33	81,81

1981-82 I. First treatment on 3/12/81, II Second treatment on 5/12/81
1985-86 I. First treatment on 7/12/85, II Second treatment on 9/12/85

 * Percentage of the total number of mites in the beehive, fallen after the first application.
** Percentage of the remaining (after the first application) mites fallen after the second application.

(1981-82) Mean: I = 77,71%(1‰) and 81,60(0,5‰), II = 79,77%(1‰) and 75,51(0,5‰)
(1985-86) Mean: I = 78,31%(1‰) and 80,99(0,5‰), II = 79,49%(1‰) and 79,33(0,5‰)

Percentage of mites fallen after two application
with Malathion 1‰ (2grs) in 10 beehives at
Votanikos - Athens, Greece.

---

a1 : Winter 1981-82, a2 : Winter 1985-86

b1 : 1st application with Malathion 1‰

b2 : 2nd application with Malathion 1‰ (Two days later)
(Percentage of the remaining mites fallen after first
application).

	Beehive	b1	b2	Total
	1	83,24	13,42	96,66
	2	72,75	15,73	88,48
a1	3	76,76	20,71	97,47
	4	86,27	11,77	98,04
	5	69,52	26,27	95,79
	6	83,59	12,78	96,37
	7	74,47	16,94	91,41
a2	8	79,32	17,31	96,63
	9	70,84	25,58	96,42
	10	83,33	13,64	96,97

The statistical analysis has shown that the factor A is not significant (F = 0,16) and the factor B is highly significant (F = 796,10)

T A B L E   V

Percentage of mites fallen after two applications
with Malathion 0,5‰ (5 grs) in 10 beehives at
Votanikos , Athens, Greece

a1 : Winter 1981 - 82, a2 : Winter 1985-86

b1 : 1st application with Malathion 0,5‰

b2 : 2nd application with Malathion 0,5‰.(Two days later)
(Percentage of the remaining mites fallen after first
application).

	Beehive	b1	b2	Total
	1	83,20	12,39	95,59
	2	75,23	15,89	91,12
a1	3	77,39	19,55	96,94
	4	89,39	7,58	96,97
	5	82,78	14,09	96,87
	6	82,19	13,21	95,40
	7	78,62	14,71	93,33
a2	8	78,51	19,02	97,53
	9	83,48	13,77	97,25
	10	82,14	14,61	96,75

The statistical analysis has shown the factor A is not significant
($F = 0,02$) and the factor B is highly significant ($F = 281,38$).

Malathion fumigation for control of Varroatosis

M.D.Ifantidis
University of Thessaloniki, Greece
O.Van Laere, H.Ramon & L.De Wael
State Research Station for Nematology & Entomology, Merelbeke, Belgium

SUMMARY

The insecticide malathion is used in Greece since 1979 for control of Varroa jacobsoni Oud. in the form of powder with a very satisfactory efficiency. Fumigation strips with malathion were used in this paper in order to make the treatment less laborious. The results were quite satisfactory in laboratory experiments but rather disappointing in field experiments.

1. INTRODUCTION

The insecticide malathion is widely used in Greece against the Varroa mite for several years in the form of powder. For each treatment 2 g of dusting material per hive, in a concentration of 1000 ppm (SANTAS, 1983) or 30 g of the preparation (3 g per frame) with a concentration of 50 ppm of the active substances (IFANTIDIS, 1981) is recommended. In this way 99 % of the mites are killed (IFANTIDIS, 1981). The treatment is harmful to open brood and higher concentrations appear to be harmful for the bees.

Because malathion-powder is commonly used against Varroa jacobsoni in Greece - the easy application and the low price of the product are playing undoubtedly an important part - we were looking for more efficient methods with malathion.

As the use of malathion in the form of powder makes the distribution in the hive difficult, methods to distribute the product homogeneously in in the hive were looked for. Two techniques are obvious : the use of fumigation strips and the aerosol-technique.

2. MATERIALS AND METHODS

2.1. The preparation of fumigation strips

Strips of strongly absorbing paper (280 g/m^2) of 45 x 70 mm are immersed in an 8 % potassium nitrate solution until saturation.
Afterwards the strips are dried horizontally on non-absorbing plates at room temperature, during about 12 hours. In order to prevent local increases of potassium nitrate, the strips should not be piled up on each other or placed vertically while drying. Each ticket contains 40 mg potassium nitrate.

Furtheron a known amount of technical malathion (with 95 % active substance) is dissolved in 0,5 ml xylol, which is dropped on the ticket and also dried at room temperature during about 12 hours.

The drying temperature of the ticket is quite important. Three different temperatures were tested. The results are reproduced in table 1

Table 1. Mortality of bees in days after fumigating with malathion tickets dried at different temperatures

| Test object 1 ug malathion | Initial number of bees | Number of killed bees after a fumigation time of 1 and 2 days at different during temperatures | | | | | |
| | | 25°C | | 45°C | | 55°C | |
		1 d	2 d	1 d	2 d	1 d	2 d
Blanco	30	1	1	13*	17*	0	0
0 (+40mg KNO$_3$)	30	0	0	1	2	1	1
115	30	-	-	3	5	2	2
330	30	8	11	3	4	-	-
660	30	28	30	30	30	2	5
1000	30	30	30	30	30	-	-
1150	30	30	30	30	30	7	8

* The high mortality is due to lack of water

The drying at a temperature of 55°C made the tickets for the bees clearly less toxic. This was confirmed by gaschromatographical determinations of the tickets, on which 150 ug malathion was dropped and which were dried afterwards at 25 - 35 - 45 and 55°C respectively (table 2).

Table 2. Quantity of malathion in ug on tickets with originally 150 ug dried at different temperatures

Drying temperature	25°C	35°C	45°C	55°C
ug malathion	148,3	158,5	151,6	128,3

2.2. Laboratory experiments

2.2.1. Influence of malathion on bees

To test the toxicity of malathion on bees, three times ten bees were placed in a small cage. These cages were put in a perspex hive of 60,75 dm3 and treated with active smoke circulation during 15 minutes (figure 1).

The cylindrical cages have a wire gauze of 0.76 mm, so that the smoke can enter the cages unhindered. Via the cover the bees were provided with candy and water (figure 2).

Afterwards the 3 cages were placed in an incubator at 34°C. After 24 and 48 hours, the mortality is checked.

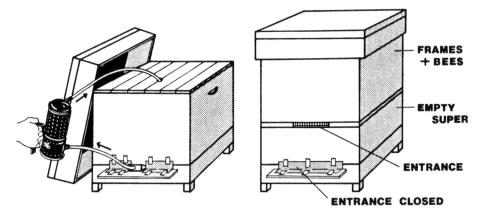

Fig. 1. Laboratory experiment in "Langstroth" brood chamber

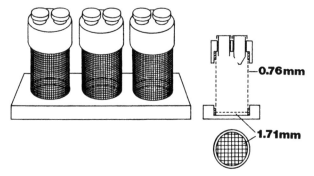

Fig. 2. The cylindrical cages for the laboratory experiments

2.2.2. Influence of malathion on bees and mites

Via the same experimental set the effect on bees and mites with maximum concentration (safe for the bees) was checked. For this purpose ten bees were placed in each cage with at least one Varroa-mite on each bee.

The three cylindrical cages are placed on a plastic plate of 8 mm. In this plate three cylindrical holes of 6 mm depth, in which the cages fit, were bored out. These cylindrical holes are provided with a filterpaper. As the bottom of the cages are provided with a wire gauze of 1.71 mm, the killed mites are falling down through the gauze on the filterpaper. In this way the killed bees as well as the killed mites can be counted easily.

The mites, which fell down on the filterpaper and recovered afterwards, can join easily with the bees, because the distance between the filterpaper and the gauze of the cage is hardly 1 mm (figure 2).

2.3. Field experiments (figure 3)

Broodless hives (on one body) are fumigated with active smoke circulation during 15 minutes. Afterwards an empty super is placed between the bottom and the body.

197

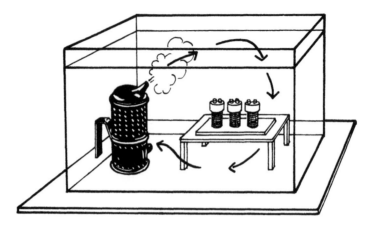

Fig. 3. Fumigating the hive and disposition after fumigating

The entrance in the bottom is closed and a new entrance is made between the super and the body. The new entrance is provided with a queen-excluder, so that killed bees on the bottom cannot be removed.

The bottom is also provided with a gauze. In this way the fallen mites are inaccessible for the bees.

After 24 and 48 hours the killed bees and the killed mites are counted.

An estimation of the number of bees is taken in function of the number of occupied frames. The percentage of infested bees before and after each treatment is obtained as follows. In each hive three samples of 150 bees were taken. For each sample three frames are partly brushed off via a funnel in a small screw-stoppered cup of 100 - 150 ml, filled half with ethanol 25 %. After 24 and 48 hours the cup is shaken and afterwards the contents of the cup are brought into a wire gauze bag, which is vigorously shaken under water in a bowl. The killed mites are falling down to the bottom and are counted.

3. RESULTS

3.1. Laboratory experiments

3.1.1. Influence of malathion on bees

After some preliminary experiments, smoke tickets of 0 - 50 - 100 - 150 - 200 and 250 ug malathion were checked. Moreover each smoke ticket contains about 40 mg potassium nitrate. The mortality of the bees after 24 and 48 hours is reproduced in figure 4.

It is clear that the concentration may not increase above 150 ug active substance of malathion. Fumigations with higher concentrations are unjustified.

3.1.2. Influence of malathion on bees + mites

In the preliminary examination bees and mites were exposed to fumigations with different concentrations. These tickets were potassium

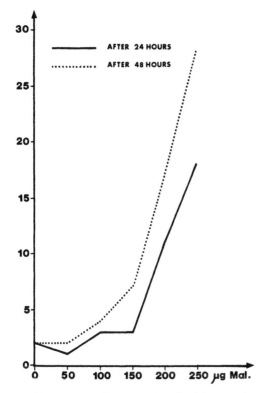

Fig. 4. Influence of malathion on bees

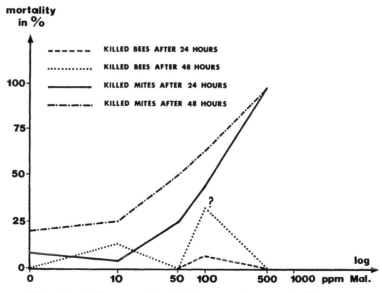

Fig. 5. Influence of malathion on bees and mites

199

Table 3. Mortality of bees and mites on 1 and 2 days after fumigating with Malathion tickets

Test object in ug malathion	Initial number of bees mites		Number of killed bees or mites after a fumigation time of				Mortality in % bees mites	
			1 day		2 days			
	bees	mites	bees	mites	bees	mites	bees	mites
0	27	28	1	0	2	1	7	4
130	27	31	1	30	1	30	4	97
160	30	30	0	28	0	28	0	93

Table 4. Mortality of bees and mites 8 and 25 hours after fumigating with malathion tickets

Test object in ug malathion	Initial number of bees mites		Number of killed bees or mites after a fumigation time of				Mortality in % bees mites	
			8 h		25 h			
	bees	mites	bees	mites	bees	mites	bees	mites
130	30	27	0	27	0	27	0	100
160	32	33	0	30	0	33	0	100

nitrate strips (cfr. above) immersed in an aqueous malathion emulsion of 10 - 50 - 100 or 500 ppm until saturation of the paper and afterwards dried at 50°C. According to gaschromatographical determinations, 500 ppm agreed with about 35 ug.

In figure 5 the influence of malathion on bees and mites is reproduced.

There was no abnormal bee mortality, except for 100 ppm, where the high mortality was due to lack of water. For comparison, Folbex VA gave 100 % mortality for the mites and 3 % for the bees, after 48 hours.

At last tickets of 130 and 160 ug malathion were checked. The results are reproduced in tables 3 and 4.

Without any risk to the bees, a mortality of the mites of 95 % is reached in the laboratory experiments.

3.2. Field experiments

The acaricide activity of malathion was very low (table 5) in these experiments.

Possible causes for this low acaricide effect are perhaps the difficult homogeneous distribution in the hive with bees, the blocage of the smoke in empty cells, the lower temperatures than in laboratory circumstances, the fact that the mites move to less accessible places when the colony is disturbed.

Table 5. Influence of fumigating with tickets of 160 ug malathion on bees
and mites in three hives

Number of the hive	Infestigation in %		Number of fallen	
	before fumigating	after fumigating	bees	mites
23	3,7	3,8	9	11
25	2,8	2,6	0	14
31	6,9	3,9	14	24

4. CONCLUSION

In the laboratory experiments, a very little amount of active matter of malathion is satisfactory to have a mortality of 95 % of the mites without any risk for the bees. On the other hand, one has to work out the application of malathion-strips in the field experiment.

When the analysis of the formed co-products shows that they are without danger to people and bees, malathion-strips are a step in the right direction for control of Varroa jacobsoni Oud. with chemicals.

Also the aerosol gives some perspectives.

ACKNOWLEDGEMENT

The authors gratefully acknowledge the skillful technical assistance of Mr. Johan DICK in the experiments.

REFERENCES

IFANTIDIS M. (1980). Malathion als kontaktmittel zur bekämpfung der Varroa-milbe. pg. 144-149. In : Diagnose und therapie der Varroatose. Bukarest, Romania : Apimondia publishing house.

SANTAS L.A. (1983). Varroa disease in Greece and its control with malathion. In : Varroa jacobsoni Oud. affecting honey bees : Present Status and Needs. Proceedings of the Meeting of the E.C. Experts Group/Wageningen, 7-9 February 1983.

VAN LAERE O., IFANTIDIS M. and DE WAEL L. (1983). Dicofol-räuchern von honigbienen zur bekämpfung der milbe Varroa jacobsoni. Apidologie 1983 14 (3), 175-182.

Effectiveness of aqueous solution of Malathion against Varroa mite applied in field experiments

M.D.Ifantidis, A.Thrasyvoulou & N.Pappas
Laboratory of Apiculture-Sericulture, School of Agriculture, Aristotle University, Thessaloniki, Greece

Abstract

The effectiveness of aqueous solution of malathion 5 ppm was tested against Varroa mite under field conditions in both nuclei and Langstroth hives. Honeybees of nuclei were kept confined during the experiment and there after they were killed, to estimate the total number of parasites. In the case of Langstroth hives, honeybees were not confined and only a part of them was killed after the end of the experiment. One treatment with the above preparation was enough to kill 99% of mites in nuclei while a second one was necessary to reach the same level of mortality of mites in standard hives. On the other hand malathion can be kept dissolved in different organic solvents at least for 6 months without losing of its effectiveness against the Varroa mite. The significant of these findings into applied beekeeping is discussed.

Introduction

Malathion has been used extensively against Varroa mite in Greece since 1979. It's effectiveness was studied by Ifantidis (1981), Pelekasis et al., (1981), Santas (1983) and Tselios and Kostarelou-Damianidou (1984). Studies conserning malathion residues in greek honeys were also made by Thrasyvoulou et al., (1985, 1986).

In Greece, malathion was registered against Varroa mite as dust at a concentration of 50 ppm and of 1000 ppm in 1985. In the first case 30 g of dusting material is recommented to dust all the bees of a colony (Ifantidis 1981) while in the second one 2 g of dusting material is recommented to be applied in the bee-space (Santas, 1983).

In laboratory tests aqueous solution of malathion was found absolutely toxic against Varroa at a concentration of 7 ppm (Ifantidis, 1981). In order to transfer this form of treatment into applied beekeeping and also to reduce the pesticide residues within the hive as low as possible we tested the effectiveness of malathion in aqueous solution of 5 ppm against Varroa mite under field conditions.

Material and methods

Aqueous solutions of 5 ppm malathion were prepared the day of the treatment by diluting the necessary amount of the pure chemical first in a negligable amount of ethyl alcohol and then, in the appropriate volume of water.

The preparation was applied on the adult bees of broodless colonies in 7 small hives (nuclei) and in 7 Langstroth hives during autumn. Four nuclei and 7 Langstroth hives were used as control.

Table 1. Effectiveness of 5 ppm aqueous solution of malathion against _Varroa_ mite applied in broodless nucleus.

Treatment	Nucleus No	Initial number of mites*	Dead mites after the application			Survivor mites
			24 h	48 h	72 h	
Malathion 5 ppm	1	379	364 (96%)	8 (2%)	2 (1%)	5 (1%)
	2	239	215 (90%)	20 (8%)	2 (1%)	2 (1%)
	3	261	241 (92%)	10 (4%)	7 (3%)	3 (1%)
	4	292	288 (99%)	3 (1%)	0 (0%)	1 (0%)
	5	219	217 (99%)	0 (0%)	0 (0%)	2 (1%)
	6	425	411 (97%)	10 (2%)	0 (0%)	4 (1%)
	7	318	309 (97%)	4 (1%)	2 (1%)	3 (1%)
	Total	2133	2045 (95%)	55 (3%)	13 (1%)	20 (1%)
Control (water)	1	291	11 (4%)	4 (1%)	8 (3%)	268 (92%)
	2	233	5 (2%)	3 (1%)	4 (2%)	221 (95%)
	3	410	2 (0%)	3 (1%)	2 (0%)	403 (98%)
	4	322	2 (1%)	2 (1%)	3 (1%)	315 (98%)
	Total	1256	20 (2%)	12 (1%)	17 (1%)	1207 (96%)

* It was estimated by adding the dead mites to the survivals that had been found on bee population.

Each frame was separately sprayed until the bees became totally wet. Aproximately, 25 ml solution was used for each standard frame. Dead mites and bees were trapped within the hives according to the method described by Ifantidis (1981).

Each of the nuclei had three frames. Their bees were confined the night before treatment and kept closed in a well airing place until the end of the experiment. They received a single treatment. Observations for dead mites and bees were taken 24, 48 and 72 hours after treatment. The total number of mites in each nucleus was estimated by adding up the number of mites that were found dead in the traps to those that survived. The latters were found by killing the whole bee population in 25% ethyl alcohol at the end of the experiment as described by De Jong et al., (1982).

In Langstroth hives the bee population covered about 10 frames.The flying of the bees was free before and after treatment. The number of mites of each colony was determined as follows.Three samples ,each of about 100 bees ,were taken from three different frames of every experimental colony. The collected bees were then killed in 25% aqueous solution of ethyl alcohol and the mites that they carried on were found and counted on the bottom of a washbowl.The percentage of the infestation from each colony was converted into the number of mites that were parasitized on the bees of each colony.The number of bees

Table 2. Effectiveness of 5 ppm aqueous solution of malathion against Varroa mites applied in broodless Langstroth hives.

| Treatment | Hives No | Initial number of mites* | Dead mites after the | | Survivors mites |
			1st treatment	2nd treatment	
Malathion 5 ppm	1	1302	1005 (77%)	284 (22%)	13 (1%)
	2	1421	1298 (91%)	120 (8%)	3 (0%)
	3	237	185 (78%)	52 (22%)	0 (0%)
	4	439	394 (90%)	44 (10%)	1 (0%)
	5	346	262 (76%)	84 (24%)	0 (0%)
	6	155	131 (85%)	24 (15%)	0 (0%)
	7	88	76 (86%)	12 (14%)	0 (0%)
	Total	3988	3351 (84%)	620 (15,5%)	17(0,5%)
Control (water)	1	1566	17 (1%)	45 (3%)	1504 (96%)
	2	450	9 (2%)	15 (3%)	426 (95%)
	3	550	3 (1%)	3 (1%)	544 (99%)
	4	306	3 (1%)	2 (1%)	301 (98%)
	5	391	2 (1%)	3 (1%)	386 (99%)
	6	215	2 (1%)	2 (1%)	211 (98%)
	7	530	1 (0%)	21 (4%)	508 (96%)
	Total	4008	37 (1%)	91 (2%)	3880 (97%)

* It was estimated by adding the dead mites to the survivors that had been found on bee population after the second treatment.

was calculated from the weight of the bee population itself.

Control colonies were sprayed with tap-water in which an amount of ethyl alcohol was dissolved as in the case of preparation of the aqueous solution of malathion 5 ppm.

To test whether the aqueous solution of malathion(·5 ppm) has residual effect on the mites, three groups of honeybees were immersed in it for one minute and then were put in incubator (34±1°C). After 24 hours 36 mites, were collected from untreated colonies and transferred on the "contaminated" bees (12 mites in each group). The same experiment was repeated with another three groups of honeybees which were sprayed with the chemical.

Since the commercial preparations of malathion may be offered to beekeepers in form of solution with in an organic solvent we tested the stability of malathion in ethyl alcohol, acetone, isooctane and hexane. For this purpose 5 mg malathion was dissolved in 5 ml of each of the above solvents (1000 ppm), placed on sealed vial and kepted at room temperature or in the refrigerator (4±2°C) six months. Analysis for malathion reduction was performed every month. Thrasyvoulou et al,1984).

Table 3. The effect of storage on the stability of malathion preserved for six months in different organic solvents.

Solvents		April	May	June	July	August	Sept.
				Concentration of malathion (ppm)*			
Ethyl alcohol	R.1000	892	748	760	888	788
	f.1000	952	768	780	1072	1056
Acetone	R.1000	920	812	800	1016	960
	f.1000	960	816	880	976	1028
Isooctane	R.1000	948	796	908	904	896
	f.1000	828	828	800	1036	937
Hexane	R.1000	920	860	840	988	920
	f.1000	948	880	840	920	896

* Initial dilution:5 mg malathion in 5 ml solvent.

R=room temperature

f=refrigerator temperature (4±1°C)

Table 4. Varroa mite toxicity of malathion kept in ethyl-alcohol solution for 6 months.

Code number of replications	LC_{50} of malathion against Varroa mite in ppm	
	before storage	After storage
1	0.320	0.100
2	0.304	0.105
3	0.147	0.228
Mean*	0.257	0.144

*Means are not significantly different in 0.05 level.

The LC_{50} of malathion against Varroa mite was tested before and after six months storage. For this purpose mites were completely immersed in the different concentrations of malathion for one minute. Subsequently, they were strained on filter paper and transferred on untreated bees into clean recovery cages in an incubator (34±1°C).The number of dead mites was recorded 48 h after the application of the pesticide and LC_{50} was estimated by calculating the regression line relating probit and log dose as discribed by Busvine (1971).

Results and discussion

Table 1 shows the effect of aqueous solution of malathion(5 ppm)against Varroa parasitizing on bees in nuclei. Since the bees were not allowed to fly and the survivor mites were also counted at the end of the experiment, the absolute number of parasites in each nucleus was estimated.From the initial population (2133), about 95% (2045) of mites were killed within the first 24 hours. The percentage of dead mites, exceeded 90% in all nuclei in this period while the coresponding percentage remained below of 5% in the controls in the first as well as in the next days. (Table 1).

Table 2 shows the effect of malathion solution of 5ppm against the mites parasitizing on bee colonies in Langstroth hives. Out of 3988 mites from all colonies 3351(84%) were killed after the first treatment. The percentage of dead mites in this treatment ranged from 76% to 91%. The relatively lower effectiveness of the preparation in this case could be attributed to that number of mites that survived probably because they were on foragers bees in the field during the treatment. After the second treatment an additional number of mites were killed so that the final percentage of dead mites was risen to 99%. Thus, a second treatment, three days after the first one,could be proved useful in cases that bees could not be confined or the application could not be applied at noon when the bees are not flying.

With laboratory tests it was found that the aqueous solution of malathion 5 ppm has no residual effect on the mites 24h after the application of the acaricide.

Table 3 shows the results of preservation of malathion that was dissolved in 4 organic solvents and kept for 6 months at room temperature or in the refrigerator. In all cases malathion remained in high concentration and no indications existed that,it was considerably reduced during storage in these solvents.

Table 4 shows the results of biossay on the lethal concentration of malathion that had been dissolved in ethyl alcohol and stored for six (6) months. No significant differences were found between the application before and after storage. This is a further indication that malathion remains rather stable in organic solvents during storage.

The level of honey contamination with aqueous solution of malathion is the subject of a paper in preparation.

The separately spray of frame by frame is somehow time consuming and this could be a disadvantage for the professional beekeeper who has several hundred hives. Because of this he may be unwilling to adopt this method.In that case the apparatus "Micro-Diffuseur S" of the "Société Phagogene" that allows to spray the bees without removing the frames within 2 minutes (Colin et al., 1983) can be very useful.

For the beekeeper who had several hives and the time is not a limited factor the spray of aqueous solution of malathion could be a valuable method for controlling Varroa mite. Another positive aspect of this method is that during treatment honeybees become calm and stings are significantly restricted.

The fact that malathion remains stable in organic solvents may help to commercialize the chemical at the right concentrations.

BIBLIOGRAPHY

1. Busvine J.R. 1971. Toxicological statistics 263-288.In Critical Review of the Techniques for Testing Insecticides. Commonwealth Institute of Entomology. England p. 345.

2. Colin M.E., J.P. Faucon and M. Morand 1983. Utilisation of aerosol to treat bee-colonies against Varroatosis (Varroase) 71-72. In Varroa jacobsoni Oud. Affecting Honey Bees: Present status and Needs. Edited by R. Cavalloro. Commission of the European Communities. A.A. Balkema Rotterdam p. 107.

3. De Jong, D. ;De Andrea ,Rona and Goncalves,L.S. 1982 . A comparative analysis of shaking solutions for the detection of Varroa jacobsoni on adult honey bees Apidologie 13 (3):297-306.

4. Ifantidis,M.D. 1981. Malathion als Kontaktmittel zur Bekämpfung der Varroa - Milbe 144-49. In Diagnose und Therapie der Varroatose, Buckarest Romania. Apimondia, Publishing house.

5. Pelekassis, C.D., L.A. Santas, N.E. Emmanuel 1981. Vorläufige Untersuchungen über die Wirksamkeit einer Malathion-Behandlung und einer Einrichtung zur langsamen Abgabe von SO_2 gegen die Varroatose in Griechenland. Int. Symp. Diagnose und Therapie der Varroatose. Oberursel-Bad Homburg. Apimondia Verlag. Buckarest, 127-129.

6. Santas,L. A. 1983. Varroa disease in Greece and its control with malathion. In Varroa jacobsoni Oud. affecting honey bees:Present Status and Needs. Proceedings of a Meeting of the E.C. Experts Group/ Wageningen, 7-9 February 1983.

7. Thrasyvoulou,A.T.; M.D. Ifantidis; N.L. Pappas;K. Simmons 1985. Malathion residues in Greek honey. Apidologie 16 (1):89-94.

8. Thrasyvoulou A.T.; M.D. Ifantidis; N. L. Pappas 1986. Contamination of honey and wax from malathion and coumaphos that are used against Varroa mite.J. Apic. Res. (25):in press.

9. Tselios,D. and M. Kostarelou - Damianidou 1984. Testing different acaricides against Varroa (V. jacobsoni). Agric. Research 8:169-175. in Greek .

Field observations on the trophallaxis of *Apis mellifera ligustica* Spin., using a systemic chemical product

M.Accorti
Istituto Sperimentale per la Zoologia Agraria, sez. Apicoltura, Roma, Italy

Summary

The food distribution behaviour of Apis mellifera ligustica Spin. is studied using the systemic chemical product Perizin liquid. The distribution of the acaricide in hive is tested by determining the number of fallen Braula coeca Nietzsch and Varroa jacobsoni Oud. during 186 hours. The results show a high activity level in A.m.ligustica depending on multiple secondary food exchanges.This experiment,carried out in a field trial,is in agreement with the laboratory tests of other Authors.

1.1 Materials and method

The use of systemic chemicals to treat varroatosis shows the importance of food exchanges in a bee family. Efficient trophallaxis is therefore decisive in eliminating mites (10).
In a field trial using Perizin liquid (Bayer) in Rome in march 1986, observations were made on the use of the product for the purpose of verifying the timing and modes of trophallaxis in Apis mellifera ligustica Spin.
   A water emulsion of Perizin was sprayed through a syringe in the spaces between the combs of the hives in the dosage recommended by the manufacturer,both fully concentrated and in 50% dilutions in four lots, for a total of 18 hives subdivided according to the strength of the family(1,7) (tab.I).
The fall curves of Varroa jacobsoni Oud. and Braula coeca Nietzsch were recorded at staggered intervals during the first 7 days of treatment, for a total of 13 observations,counting the parasites collected from inserts. Since it was beyond the scope of this trial to consider the efficiency of the product used,the purpose being to study the modes and timing of distribution and diffusion of food in the hive,it was considered sufficient to refer the total cumulative values of the first treatment (tab.II).This is also because there were no particularly significant differences in the fall curves for each different lot,and in any case both the doses and the concentrations used had been proved suitable for this specific purpose already in previous experiments (10).
   Taking into account the sensitivity of the lice to the product used , it was considered that by observing their fall curve it would be possible

LOT	No. of families	Average No. of bees	Concentration	Dose (cc)
I 1	5	10,000	1:100	50
I 2	4	3,000	1:100	25
II 1	5	13,300	1:50	50
II 2	4	2,900	1:50	25

tab.I

			No. of hours after treatment									
1	2	3	6	12	24	36	48	72	96	120	144	168

Braula

66.4	72.8	76.6	85.2	90.1	92.7	92.7	92.7	94.3	96.5	97.8	99.3	100

Varroa

17.3	37.7	54.1	75.7	90.7	94.0	95.2	95.8	96.9	97.8	98.2	99.1	100

tab.II: cumulative percentage of deaths of Braula and Varroa .

to determine the speed at which the product was distributed in the hive regardless of the number and frequency of trophallaxis taking place.
On the other hand, since the number of mite deaths was a function of the lethal dose of the active ingredient used,it was possible to determine indirectly from the relative death rate curve the percentage of bees involved in repeated food exchanges.
However,it is be stressed that these observations are based on two assumptions:

    1) that the parasites populations are equally distributed on the hosts;

    2) that multiple parasitization,more parasites on the same bee,is a negligible phenomenon (14).

Since the experiment in question was carried out in the field,it was difficult to identify all the variables wich in some way might have affected the trophallaxis . It is therefore considered reasonable to accept the results obtained in an essentially descriptive logic.

1.2 Results and conclusion

From the death rate curve of lice (fig.1),it is observed that after 3 to 6 hours the trophallaxis involves between 76% and 86% of the bees.
From the number of fallen mites,it was observed that in the same time period the repeated trophallaxis involved between 54% and 76% of the bees concerned.
The difference between both the trends is presumably due to the fact that food exchanges vary according to the sex,age and caste of the individuals in question (3,6,9), and that there must be numerous exchanges of food

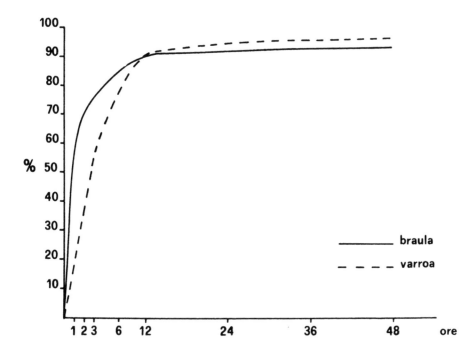

Fig.1 :Cumulative percentage of deaths of Braula and Varroa.

before the active ingredient of Perizin liquid reaches the lethal concentration for the mites in the haemolymph of the bees(12).
This rate of food exchange and distribution in the hive has already been
encountered by other Authors and it appears to be a characteristic of the
different species and races of bees(4,5,11).
     Since the results are in agreement with :
1)The laboratory tests on trophallaxis;
2)the effectiveness of Perizin liquid in the control of varroatosis(8,13);
3)the death rate curve of lice and mites obtained in Italy(8),
we think that it is negligible where there is also a contact or inhalation
action of Perizin.Moreover ,at present time,both supposition are not yet
proved.
To conclude, despite all the possibly observable uncertainties,the information obtained from the field trial leads us to accept the fact that
within 12 hours of giving the Perizin liquid,there is a uniform distribution of the food as shown by the 90% level of deaths in the cumulative
fall curves of the lice and mites.
This result is not substantially different from the estimated 24-48
hours needed to treat the entire (100%) bee population (2,11).

REFERENCES

1. ACCORTI, M.(1985). Valutazione numerica degli adulti di Apis mellifera
   L.:variazioni e modifiche al metodo dei sesti.Apicoltura 1:63-73.
2. BOLTEN A.B.,ROBINSON F.A.,NATION J.L.,YU S.J.(1983).Food sharing
   between honeybee colonies in flight cages . J.Apic.Res.22 (2):98-100.
3. GALLIOT G.,MONTAGNER H., AZOEUF P. (1962).Etude quantitatives des
   transferts de nourriture entre ouvrieres et males chez l'abeille
   domestique (Apis mellifica L.). Insectes Sociaux 29 (2):268-279.
4. KLOFT W.J.,DJALAL A.S.,DRESCHER W. (1976).Untersuchung der unter-
   schiedlichen Futterverteilung in Arbeiterinnengruppen verschiedener
   Rassen von Apis mellifica L. mit hilfe von P^{32} als Tracer.Apidologie
   7 (1):49-60.
5. KLOFT W.J.,ROBINSON F.A. (1976).Food sharing behaviour among caged
   workers of different races and species of honey bees.Am.Bee J.116(3):
   106-107.
6. KORST P.J.A.M.,VELTHUIS H.H.W. (1982).The nature the trophallaxis in
   honeybees. Insectes Sociaux 29 (2):209-221.
7. MARCHETTI S. (1985).Il "Metodo dei sesti" per la valutazione numerica
   degli adulti in famiglie di Apis mellifera L. .Apicoltura 1:41-61.
8. MARCHETTI S.,D'AGARO M. (1986).Perizin liquid,a systemic agent for
   the chemical control of varroatosis.Apicoltura 2:in press.
9. MONTGNER H.,PAIN J. (1971).Etude preliminaire des communications entre
   ouvriéres d'abeilles au cours de la trophallaxie.Insectes Sociaux
   18 (3):177-192.
10. MORITZ R.F.A.,KOENIGER N.,MAUL V. (1981).Verteilung systemisch wirken-
    der Präparat im Bienenvolk (Apis mellifera L.).Diagnose und Therapie
    der Varroatose.Int.Symp.Bienenbiologie und -pathologie.Oberursel,
    29/IX-1/X 1980:25-38.
11. NIXON H.L.,RIBBANDS C.R. (1952).Food trasmission in the honeybee
    community.Proc.ent.Soc.Lond. B 140:43-50.
12. PERSHAD S. (1967).Analyse de différents facteurs conditionant les
    échanges alimentaires dans une colonie d'abeilles Apis mellifica L.
    au moyen du radio-isotope P^{32} .Ann.Abeille 10 (3):139-197.
13. RITTER W. (1986).Versuche zur Entwicklung und Prüfung von Perizin,
    einem systemischen Medikament zur Bekämpfung der Varroatose der
    Honigbiene. Tierarztliche Umschau 41 (2):105-112.
14. SMITH I.B.,CARON D.M. (1984).Distribution of the bee louse Braula
    coeca Nietzsch in honeybee colonies and its preferences among workers,
    qeens and drones. J.Apic.Res. 23 (3):171-176.

Different modes of Perizin® application

A.Klepsch
Hessische Landesanstalt für Leistungsprüfungen in der Tierzucht Neu-Ulrichstein, Abteilung für Bienenzucht, Kirchhain, FR Germany

Summary

Perizin™, a new drug for Varroa control was registered in December 1985. This new drug should be used during late autumn or early winter. Only little field data were available on treatment of bee colonies with Perizin™ in early spring. On 11.3. and 18.3., Perizin™ was applicated to 13 colonies. 99.2% of the mites had been killed. Another experiment showed that the bee colonies tolerate the treatment. Later on we tried treatment of artifical swarms. A method was developed by which more than 95% of the mites can be killed with a single application of Perizin™. This method is using the equipment and dosage as provided within the registration.

1.1 Introduction

1982 in Hessen regular control of Varroatosis was started. From the very beginning we tried to have a synchronos treatment for all infected bee colonies. From late autumn to early spring timing does not need to be very accurat, because the risk of reinfection is very low while bees do not fly. Also most medications do only hit those mites, which are not protected inside the sealed brood. Both reasons together show that the best time for treatment will be during this period and should be started as early as possible to get rid of the mites.

Perizin™ was registered for Bayer AG in December 1985. The regular treatment of Varroatosis should be finished at this time. In Hessen, however, the preceding regular treatment with formic acid turned out to be rather inefficient. To avoid damages to the bees it was necessary, therefore, to add one more treatment in spring. For this purpose Perizin™ was chosen. Since only very little data were available on treatment in spring, we started researches on efficency of Perizin™ against Varroa and also we studied how the bees will stand this treatment.

From the very beginning of regular and organized Varroa control, we had to learn that in many cases one treatment a year may not be enough to avoid damages to any infected bee colony, even when medications are used which will kill nearly 100% of the mites inside a colony. So it is important for the beekeepers to have methods for Varroa control they can use in addition to the regular treatment of their colonies. A newly introduced drug may offer some more chances than announced in the first registration. So we tried to find a method to cure swarms and artifical swarms with Perizin™.

213

2.1 Varroa control by Perizin™ in early spring

2.1.1 Material and methods

In order to see, if a treatment with Perizin™ has any harmful side-effect to bee colonies, for control dead bees were collected on the bottom of the hive of 145 untreated colonies in 9 apiaries between 4.mar. and 14.mar.1986. Another 45 colonies in 4 different locations were treated with 50ml of a water solution containing 1ml Perizin™ two times, on 7.mar. and on 14.mar.1986. The dead bees on the bottom were collected and counted on the first, the third and the seventh day after each treatment.

For evaluation of Varroa control in spring with Perizin™, 13 colonies in different types of hives were brought to one apiary. The treatment was performed on 11.mar. and 18.mar.1986. Inside the hives there was an insert protected with gaze in order to collect the dead mites for counting. On the 24.mar.1986 all the 13 colonies were killed. The dead bees were washed out to find the number of residual mites. At this time of the year, it was also necessary to inspect a greater number of brood cells for residual mites.

2.1.2 Results

Table (I) shows the number of dead bees per colony and day in 9 apiaries with untreated colonies. Table (I) also illustrates the number of dead bees per day in colonies after the treatment with Perizin™. The total number of dead bees per day found during the above mentioned period of a fortnight is 2.5 times higher than in the control colonies. A closer look to the data shows that most of the bees are dying immediately after treatment (table II). Most of those dead bees found after 24 hours were black and wet. The next controls nearly show the same amounts of dead bees on the bottom of the hives as in the control group (fig.I).

In March sealed brood is present in every colony (table III). Fortunately the infestation of the first brood in spring is very low (3). So only very few mites could escape from the treatment. Therefore there is a strong effect on the mites. 99.2% (min:. 94%; max.: 99.9%) of the mites were killed by two applications of Perizin™(table IV). During the registration in each of the both groups, treated and untreated colonies, one queen was lost.

2.2 Treatment of artifical swarms

2.2.1 Material and methods

Artifical swarms of a weight of ca. 2 kg were shaken from heavily infected colonies into swarm boxes. A queen was added to every swarm. The boxes were stored for 24 hours in a cool and dark place. No food was given to the bees during this time. After that time, the application of the medication or the placebo was done. We made 4 groups for treatment:
1. 1ml Perizin™ in 49ml water - 5 swarms
2. 1ml Perizin™ in 49ml syrup - 6 swarms
3.   50ml water - 2 swarms
4.   50ml syrup - 3 swarms

The mode of application had to be topical. Therefore, swarm boxes with a removeable cover were used. The box containing a swarm had to be shocked once heavily to the floor. Immediately afterwards, all the bees were laying puzzled on the bottom of the box. Then there was time enough to remove the cover from the box and to pour 50ml fluid from a beaker onto the swarm. The application took less time than one minute. Afterwards, the swarm received food ad libitum. It was stored for three more days, until the bees were killed to be washed out for residual mites of Varroa Jacobsoni. During this time, the dead mites were collected on a white plastic insert under the screened bottom of the swarm box.

2.2.2 Results

As table (V) shows, both modes of application of Perizin™ will eliminate ca. 95% of the mites. The control treatments have no effect to the mites. Less than 0.5% of the mites are dying after such a treatment. There is no significant difference between group 1 and 2 and also no significant difference between group 3 and 4. No queen was lost during all these experiments.

3. Discussion

The new drug, Perizin™, allows more ways of succesful application than there are provided within the first registration. Succesful treatment of full colonies is possible until March, even when the colonies are already rearing brood. However, the number of dead bees per day will increase immediately after application. Although the colonies will easily tolerate these losses, late treatment only should be recommended in cases of emergency. Regular Varroa control with a drug that only eliminates the mites on the bees, has to be as early as possible during a period, when bees do not fly and when there is no opportunity of reproduction for the mites. Therefore, the best time for regular Varroa control, which covers all apiaries in an area, will be in late November, if Perizin™ is used.

Foundation of new colonies should include Varroa control. As long as medications are not able to hit mites during their reproduction inside the sealed brood cells, sealed brood may not be inside the nuclei. Swarms and artifical swarms will be a good way to found new colonies. Treatment of a swarm with Perizin™ is very easy, efficent and does not disturb the regular handling of the swarm. A single application of Perizin™ will be enough. Even when not all the mites are killed, the new colonies will reach the next Varroa control without damages caused by Varroatosis. Also if more mites would be eliminated in those nuclei, they have to be treated in the next winter, because there is always a great danger of reinfection. For that reason it does not make sense to improve the efficency of the treatment by adding more applications. An improvment in the way of applying Perizin™ will be possible if topical application is left. It will be much easier for the beekeeper to add an systemical drug to the food of a swarm. Perizin™ might be usable for such a treatment.

References

1. MAUL, V., KLEPSCH, A., WACHENDöRFER, G. (1983). Befallsstärke der Varroatose in Hessen. die biene 119, 197-199
2. MAUL, V., KLEPSCH, A., WACHENDöRFER, G. (1984). Befallsstärke der Varroatose in Hessen im Herbst 1983 und Verlauf der Befallsentwicklung seit 1982. die biene 120, 489-494
3. MAUL, V., SCHNEIDER, H. (1980). Vergleichende Prüfung der Wirksamkeit von Dicofol und Phenothiazin auf Varroa jacobsoni im Feldversuch. Hessische Landesanstalt für Leistungsprüfungen in der Tierzucht Neu-Ulrichstein 1979 61-62
4. RITTER, W. (1986). Bekämpfung der Varroatose mit Perizin, einem neu zugelassenen Medikament. Allgemeine Deutsche Imkerzeitung 29, 42-43
5. RITTER, W. et al. (1986). Versuche zur Entwicklung und Prüfung von Perizin, einem systemischen Medikament zur Bekämpfung der Varroatose. Allgemeine Deutsche Imkerzeitung 29, 78-82

Table I: Dead bees/day on the bottom of hives in March 1986

a) Colonies without any treatment

Apiary	No. of colonies	dead bees/day
5	23	10 (s = 4.6)
6	11	15 (s = 5.7)
7	13	10 (s = 5.5)
8	9	8 (s = 4.4)
9	20	13 (s = 9.2)
10	15	13 (s = 8.6)
11	14	11 (s = 8.0)
12	9	14 (s = 17.2)
13	16	16 (s = 14.3)
total	149	12 (s = 8.9)

b) Colonies with two times application of Perizin
Registration started with the first application
and ended 7 days after the second application

Apiary	No. of colonies	dead bees/day
1	17	30.9 (s = 14.7)
2	11	20.7 (s = 5.7)
3	7	34.4 (s = 21.3)
4	9	33.1 (s = 19.5)
total	44	29.3 (s = 15.8)

217

Table II: Dead bees/day after application of Perizin

Apiary	No. of colonies	First treatment			Second treatment		
		No. of dead bees (standard deviation)			No. of bees (standard deviation)		
		1st day	2nd+3rd day	4th – 7th day	8th day	9th + 10th day	11th – 14th day
1	17	199.0 (115.9)	9.9 (7.6)	3.4 (2.8)	76.7 (49.9)	39.8 (34.8)	10.9 (19.3)
2	11	84.5 (29.9)	8.6 (3.1)	5.5 (3.5)	97.1 (50.6)	19.8 (10.3)	7.3 (6.7)
3	7	217.3 (126.5)	12.4 (8.8)	6.8 (6.2)	64.9 (23.3)	18.6 (10.9)	27.7 (39.0)
4	9	170.0 (79.8)	21.0 (17.1)	15.1 (17.2)	73.1 (45.2)	26.7 (18.8)	16.0 (26.1)
total	44	167.3 (105.8)	12.3 (10.5)	6.9 (9.2)	79.2 (45.9)	28.8 (25.4)	13.7 (23.11)

Table III: Distribution of mites between sealed brood and adult bees at the end of the experiment

colony No.	weight of bees (g)	No. of mites on the bees	mites/ 100 g bees	number of sealed cells	mites in cells	mites/ 100 cells
B 1	471	1	0.2	120	0	0
B 2	166	7	4.2	1000	5	0.5
B 3	535	0	0	430	5	1.2
C 5	1432	4	0.3	2700	2	0.1
C 7	1660	4	0.2	3100	3	0.1
C 10	1547	18	1.2	760	4	0.5
6	1482	1	0.1	1400	0	0
10	1513	19	1.3	5700	5	0.1
11 N	1101	4	0.4	4700	3	0.1
13	1063	3	0.3	2900	0	0
14	975	7	0.7	3500	2	0.1
62	824	2	0.2	120	0	0
11 OU	240	0	0	400	4	1

Table IV: Efficiency of Varroa control by Perizin in March 1986
Application at 11. and 18.mar.86 killing of experimental colonies on 24.mar.86

colony No.	dead mites after 1. treatment	dead mites after 2. treatment	dead mites	residual mites	efficiency %
"Blätterstöcke" B 1	479	41	520	1	99.8
B 2	693	21	714	12	98.3
B 3	706	41	747	5	99.3
Magazines C 5	409	21	430	6	98.6
C 7	1780	8	1788	7	99.6
C10	3077	98	3175	22	99.3
6	177	13	190	1	99.5
10	331	48	379	24	94.0
11 N	358	12	370	7	98.1
13	313	18	331	3	99.1
14	236	20	256	9	96.6
62	1852	197	2049	2	99.9
11 OU	1181	32	1213	4	99.7
total	11592	570	12162	103	99.2

(min. 94.0; max. 99.9)

Table V: Treatment of artificial swarms with Perizin

mode of treatment	No.	weight of swarm bees (g)	Dead mites a day after treatment	residual mites	efficiency of treatment (%)
Perizin in 50 ml H$_2$O	1	1627	1954	54	97.3
	2	2313	5133	25	99.5
	3	1712	3230	26	99.2
	4	1802	6686	371	94.7
	5	1743	3931	380	91.2
total			20934	856	96.1
Perizin in 50 ml syrup (3:2)	1	1684	1190	158	88.3
	2	1880	1431	107	93.0
	3	350	188	8	95.9
	4	1828	464	1	99.8
	5	1948	612	9	98.6
	6	1417	1414	0	100
total			5299	283	94.9
control: a) water	1	1156	15	3572	0.4
	2	1731	20	3805	0.5
total a			35	7377	0.5
b) syrup	3	1859	4	1633	0.2
	4	1871	3	1300	0.2
	5	2109	1	799	0.1
total b			8	3732	0.2
total a+b			43	11109	0.4

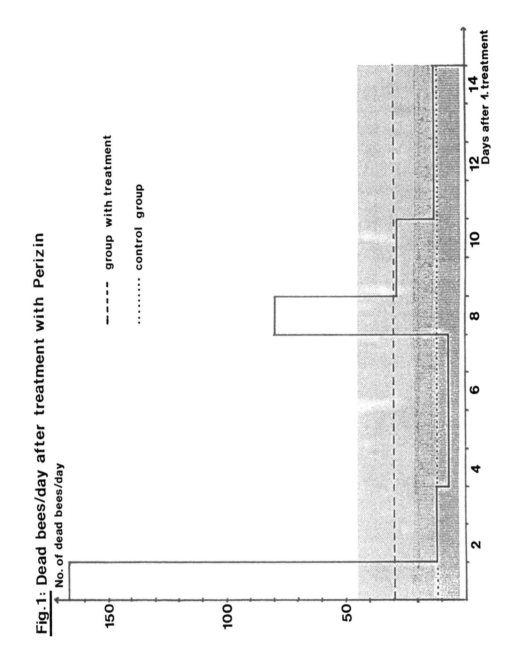

Fig.1: Dead bees/day after treatment with Perizin

No. of dead bees/day

– – – – – group with treatment

·········· control group

Days after 1. treatment

Fluvalinate, an interesting molecule against *Varroa jacobsoni*

R.Borneck
Institut Technique de l'Apiculture, Bures/Yvette, France

Summary

A numerous series of tests have been made to determine the toxicity for the bees and brood of the fluvalinate, a synthetic pyrethroid used in plant -protection for its acaricide and insecticide properties. Trying this product against Varroa Jacobsonii in field conditions and with a dosage of 1.2µg active matter per hive all varroas were dead in the hives after 8 days following the treatement. Other tests are actually taking place either in the laboratory or in the field, including a country-wide experiment on some 100.000 colonies.The results are particularly promising- The fluvalinate seems to be intersting not only for the control of Varroa but at the same time for Braula Coeca Acarapis Woodi and at a lesser extent for Galleria Mellonella larvae.

1.1 Toxicity on bees

The toxicity on bees of fluvalinate has been tested first for adverse effects on bees in the field of plant-protection. The first tests have been conducted by Joansen and Atkins in 1979 and 1980 with fluvalinate racemic (4 diastereoisomeres) and its formulation MAVRIK 2 E. In 1980, tests with small cages took place at the french I.N.R.A. laboratory in Bures -sur - Yvette, then other tests were made by Norman Gary in Davies (U.S.A.). All the results were concordant: : the acute toxicity on bees gives a topical LD 50 of 65.85 µg, an oral LD 50 of more than 200 µg, per bee. This molecule is 1 800 times less toxic than cypermethrine and 138 times less toxic than tralomethrine which is the next less toxic pyrethroideof the existing range.

Verdict : Fluvalinate, pyrethroid produced by ZOECON is non toxic for the bees. The american firmZOECON has recently been acquired by SANDOZ S.A. which soon commercialised fluvalinate with two different formulations for plant-protection. (Mavrik and Klartan .) (*)

An explanation of this surprising non-toxicity for the bees is put forward by Dr. PLAPP junior of Texas University during the annual meeting of the Entomology Society of America in 1983. - The proteines of the nervous receivers of the bees do not link with fluvalinate contrary to what happens with many others insects. This non-toxicity is shown again in France during the building up of the technical dossier presented before the homologation commission ruling the provisionnal authorisation for plant-protection products with fluvalinate in 1984. The results are shown on table 1 and 2 .

The Technical Institute for beekeeping took part in these experiments and owing to the already known acaricide power of Fluvalinate, decided to try it for the control of Varroa Jacobsonii. After some laboratory tests to make sure that the acaricide characteristics of Fluvalinate could be

extended to Varroa, other tests on dosage were undertaken to determine
the quantity of active matter the bee-colonies could bear in the field.

In december 1984, we made a first test with the microdifusion or
aerosol process we already were using with Amitraz. The results are shown
on TABLE 3.

You certainly will remark the insecticide effect of this product on
Braula Coeca. In aerosol and compared with Amitraz, Fluvalinate did'nt
show a significant bee-mortality.

1.2 Treatement against Varroa

A first field test has been done in august 1985 using DOZ 70 (Fluvalinate in a water formulation for orchards). The treatement consists in a
water solution of 1,2 mg active matter diluted in 50 ml of water for one
colony. It is gently poured straight on to the bees between the frames.

Results are shown on table 4.

The counting of dead Varroa had to be stopped after the second treatement because there were no more emerging bees and we could not find
anymore dead Varroa. Other field tests, at the same period, indicated that
the acaricide activity of Fluvalinate seems to last some 20 days after
treatements. In june 1986, the technical Institute studied more closely
the problem that could be raised by the larval deaths during direct application. We used then the B.L.M.T (Bee larval morphogenic tests) of
Dr Atkins. The chosen doses of active matter were three times those used
in the fields. The Fluvalinate leads to no larval toxicity. Other tests
conducted recently in another way gave us the same answer.

Other experiments are going on in collaboration with the firm ZOECON
to produce a veterinary speciality which could be used easily by the bee-
keepers.

However a country-wide test is at the moment taking place to appreciate
the use of Fluvalinate under different field conditions.

Some 100 000 bee-colonies are under experimental treatement in two
thonsand different locations but most of them in the southern part of the
country where the colonies have brood all the year around. They will be
followed up for behaviour and developpement to next spring by the bee-
keepers and the Technical Institute.

Many other experiments are already planned in several different direc-
tions, some of them being more fundamental - like counting Varroa under
the cappings with the R.M.N. technique (Nuclear magnetic Resonance) or
studying the effects of Fluvalinate on the neurosystem of the bee.

Some other to test the influence on Varroa of pest-control treatements applied on orchards with Fluvalinate through foragers bringing
back Fluvalinate to their hives highly infested by Varroa Jacobsonii.

1.3 Toxicologic aspect

1,2 mg of active matter per hive, or the double, in case of a second
treatement should not lead to very important residues problems. Analysis
are being made at the moment. The toxicologic record of Fluvalinate appears
particulary satisfactory.

1.4 Conclusion

I would dare to say that the Fluvalinate is liable to be the most
outstanding acaricide against Varroa in the course of the next few years
and probably against Acarapis Woodi as it appeared in the test we conducted in Canaries Islands, last spring.

Wether or not a phyto sanitary product can be legally used in the
future for fighting varroatose will be soon carefully studied in France;
at the same time, proposals will be forwarded to change fundamentally the
actual Bee policy inspired by O.I.E whichis, to say no more, out of date,
inefficient unrealistic.

REFERENCES

1 - BARNAVON M - BORNECK R. (1986)
    Toxicité des Insecticides sur les abeilles - APIACTA vol 2 et 3

2 - BORNECK R - MERLE B.
    Rapport d'Assemblée Générale ITAPI - Juillet 1986.
    Intoxication de larves d'abeilles avec Fluvalinate - Essai de rubans
    imprégnés de Fluvalinate sur abeilles naissantes.

(*) N.B. Later on the L.D 50 S were determined for the two commercial
        formulations.
        MAVRIK - Topical   LD 50 = 4,66 µg/bee
                 Ingestion LD 50 = 83,88 µg/bee
        KLARTAN Topical   LD 50 = 9,12 µg/bee (ATKINS 1983)

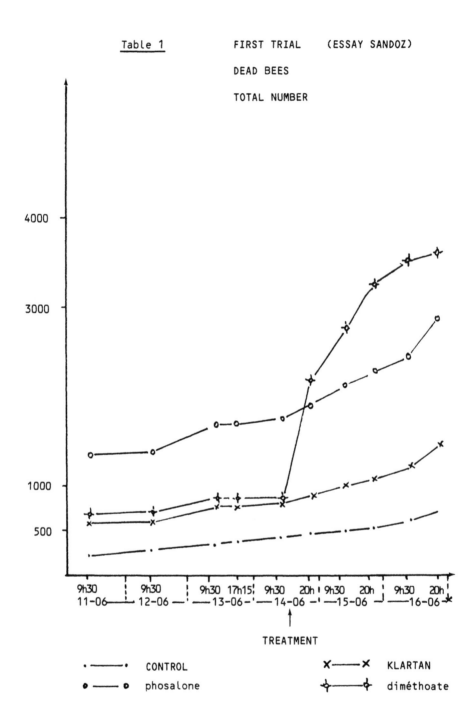

Table 1   FIRST TRIAL   (ESSAY SANDOZ)

DEAD BEES

TOTAL NUMBER

TREATMENT

•——• CONTROL   ✗——✗ KLARTAN

●——● phosalone   ✚——✚ diméthoate

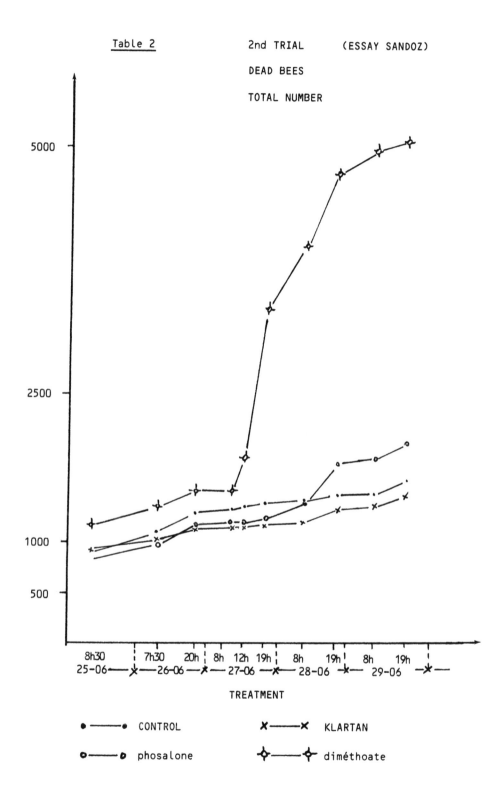

Table 2   2nd TRIAL   (ESSAY SANDOZ)

DEAD BEES

TOTAL NUMBER

TREATMENT

● —— ● CONTROL   ✗ —— ✗ KLARTAN

◒ —— ◒ phosalone   ✛ —— ✛ diméthoate

* Preliminary test   1984   FLUVALINATE-AMITRAZ   Table 3

TOXICITY on BEE-

Device : EDAR

Methodology :

Dosage : FLUVALINATE 1 200 µg A.M per colony (0,0012 g)
         AMITRAZ   24 000 µg A.M per colony (0,024 g)

Count : T + 24 h

FLUVALINATE

Number of the colonies	Dead Bees	Dead Varroas	Dead Braula Coeca
108	0	0	0
144	2	0	3
313	0	0	20
338	0	0	4
311	0	0	2
191	2	0	5
186	2	0	3
341	2	0	3
197	1	0	4
189	0	0	2

AMITRAZ

Number of the colonies	Dead Bees	Dead Varroas	Dead Braula Coeca
304	0	0	0
316	1	0	0
190	1	0	0
337	0	0	0
114	1	0	0
309	0	0	0
333	0	0	0
306	0	0	0
300	0	0	0
314	0	0	0

(*) Essay: ITAPI-CETA ALPILLESLUBERON

ESSAY FLUVALINATE AUGUST 1985   : 50 ml H2O + 0,0012 gr.A.M per HIVE

Table 4

No of HIVES	NUMBER OF FRAMES WITH BROOD	T1 Dead Varroas			T2 Dead Varroas
		J + 1	J + 5	J + 10	J + 1
350	3	603	119	98	2
64	3	357	69	71	0
68	2	182	34	28	0
92	4	369	71	92	0
95	3	1 502	200	188	1
170	3	546	102	93	3
301	2	240	48	49	0
325	4	1 809	198	206	2
84	4	1 708	203	195	3
48	4	2 252	350	342	1

T1 : FIRST TREATMENT
T2 : SECOND TREATMENT
J : DAY OF TREATMENT
J+1 : J + ONE DAY

N.B. - In August 1985 - The experimental hives have only emerging brood scattered on a few frames.

A new approach to chemotherapy of Varroatosis*

N.Koeniger
Institut für Bienenkunde (Polytechnische Gesellschaft), Fachbereich Biologie, J.W.Goethe Universität Frankfurt am Main, Oberursel, FR Germany
M.Chmielewski
Bee Diseases Research Laboratory, Agricultural University of Lublin, Lublin, Poland

SUMMARY

In traditional chemotherapy the duration of the action of the acaricide in the hive is mainly influenced by the bees and their behaviour. To increase the efficiency of the control the treatments have to be repeated. In case of FOLBEX or ILLERTISSER MILBENPLATTE (formic acid) several times. Systemic acaricides like CHLORDIMEFORM, PERIZIN and APITOL require a high load of active ingredient to be taken up initiall; by the first bees. This can be the cause of bee mortality.

We started experiments using a carrier impregnated with acaricides (pyrethroids). The release of the acaricide from the carrier remains constant over a long period. The carrier eleminates at a constant rate mites which come directly or indirectly into contact with it, until the carrier with the remaining acaride is removed from the hive.

Initial experiments with two methods, FLUVALINATE (ZOECON) foundations and wooden carriers with BAYVAROL (BAYER), are presented.

The early methods of chemotherapy of varroatosis were mainly general pest control methods. Spraying, dusting, fumigation and aerosol were typical modes of application used in the 'first generation' of control methods for Varroa jacobsoni. None of these treatments took the biological conditions in the bee colony into account.
Bees clean activly the hive from all kinds of 'pollution' and regulate the interior of the hive to a suitable and nearly constant level of temperature, humidity and CO_2. As a result of this behaviour the introduction of acaricides by the beekeeper is counteracted by the bees. So the effect of treatments is limited to the very moment of its application. At once the bees start to remove the substances. Consequently the mites are exposed to 'mortal' conditions for a limited time. All mites which are not hit during this short period survive. To increase the efficiency of the treatment it was repeated several times. For example the application of formic acid (Illertisser Milbenplatte)(Wachendörfer et al 1985) or the fumigation with FOLBEX (Klepsch et al 1983) had to be repeated several times. This led to high amounts of the applied chemicals. Residue problems in honey and bees wax are the consequences.

The further development of chemical varroa control methods brought a change in the application method. An acaricide was fed to bees and

* We want to thank the EUROPEAN COMMUNITY and the firms, ZOECON, BAYER and HAMMANN for supporting this research.

biologically distributed by natural trophallaxis in the colony. The acaricide then passed into the haemolymph of the bees and killed the mite feeding on them.

The first systemic acaricide used for vorroa control was   CHLORDIMEFORM (Ruttner et al 1980) which never was officially registered because of residue problems. Later PERIZIN   (Ritter 1985) was developed and came on the market as the 'second generation' of varoatosis treatments. Compared to the previous control methods they certainly have some advantages. The treatment of colonies is easy and fast: the syrup + acaride is sprincled into the bees ways. A single treatment is more efficient, so one or two replications are sufficient.

The systemic treatments do not 'reach' the mites in the brood. So they can be applied successfully only during autumn and winter. But winter bees are reared in July and August and in case of higher mite infestation they are already damaged during their larval and pupal development by varroa.

There is a specific problem of systemic acarides for varroa control: the bees which take up the administered syrup are 'loaded' with high amounts of acarides up to the level of toxicity. Apparently, the initial concentration has to be very high because only a limited volume can be applied and during its way from bee to bee it is diluted. So, to reach an acaricide level above varroa toxicity in a maximum number of bees some initial bee mortality has to be accepted.
Systemic acaricides are effective only a limited period after application (until they are naturally eliminated out of the bee's haemolymph). In this regard they have no advantage compared to previuos methods like Folbex etc.
Among others one point seems to be of crucial importance for improvment of chemotherapy of varroatosis. The period and length of action of the acaricide in the infested bee colony has to be at the command of the beekeeper.
It takes about 12 days until all sealed brood cells emerge and until all varroa mites are exposed to an acarides which is dispersed among the bees. So during the bee season the required mimal lenght of a treatment is 12 days. It is a major advantage if the duration of each treatment can be adjusted to the specific situation.
We tried to achieve a constant release of the acaricide within the colony by using of carriers which are introduced into the hive. These carriers were impregnated with acaricides which should cause a constant rate of mite mortality in the colony. At the end of the treatment the carrier with the acaride can be removed.

Because the life cycle of Varroa jacobsoni is naturally divided into two periods we developed 2 different methods which will be described and discussed seperately:
> 1. Carriers are placed in the colony and kill the mites on the bees which come directly or indirectly into contact with the carrier.
> 2. Impregnated wax foundations eliminate the mites which come into contact with the combs.

1. Carriers.

Carrier material.
Paper tissue, cotton or similar material normally is fairly soon
desintegrated and carried away by the bees. So with this material the
beekeeper has no control of the duration of the treatment. Further the
uncontroled disposal of the carrier together with the acaricide by the
bees may cause problems in and outside the hive.
The major advantage of carriers is that the bees are not able to remove
them. So our experiments were restricted to material which could resist
the natural cleaning behaviour and which can be removed at the end of
the treatment by the beekeeper.
The principle of dispersal of the acaride within the colony is by
contact from the carrier to the bee (and the mite on this bee) and then
from bee to bee and on the varroa mites on them. By cage tests this
pathway was demonstrated:

> Bees were exposed to a carrier with BAYVAROL in a cage for 24 h. Then
> these bees were transferred to a new cage without acaride. Some bees
> carrying a mite each were added into the new cage. The mites were
> killed within 24 h. Varroa mites survived in control tests (same
> treatment but without acaricide).

Naturally the queen substance in the colony is dispersed in a similar
way. The queen produces the pheromone which is tarnsferred by contact to
'messenger' bees which then transport it to the other bees of the colony
(Seeley 1979)

First experiments were started using carriers out of plywood and later
some synthetic material like polyester, pvc, polyurethan, polyethylen
and polystyrol was tested. The structure of the carrier and chemical
nature of the acaricide both are very important for the effect of the
release and must be very well adjusted.

Acaricides.
Our experience is limited to pyrethroids. The experiments reported here
were done with BAYVAROL, a pyrethroid produced by BAYER AG which has a
very high varroacidal effect. We used plywood carriers which were
impregnated with a dosis of 0.5 $\mu g/cm^2$ BAYVAROL.

Position in the hive.
We tested 4 different positions in a hive.
> 1. An impregnated inner cover was placed above the combs.
> 2. Two impregnated partitions were placed in the bees way between
> the combs.
> 3. An impregnated plywood grid (Bausperre) was placed under the
> combs.
> 4. A carrier was placed around the hive entrance.

In this experiment we got a good effect with the comb partitions, which
was about 99% varroa elimination. The other positions were less
efficient. Some field trials with larger numbers of colonies will be
carried out in autumn 1986.

2. Impregnated comb foundations.

a. Fabrication of wax foundations.
Wax foundations were manufactured in cooperation with HAMMANN using his
automatic processing equipment. FLUVALINATE, a pyrethroid produced by
ZOECON, was added to pure bees wax up to 0.05 % (weight) and mixed
carefully at 60°C. This liquid mixture was pumped then to rolls where
the foundations were pressed and packed.

b. Field trial.
Material and methods.
The bee colony received a super with 10 FLUVALINATE foundation in each.
5 liters of sugar syrup was fed to stimulate comb building. After a week
the queen with one 'normal' brood comb was brought to the super with the
newly built FLUVALINATE combs and separated by a queen excluder from the
previous brood chamber. The normal brood frame was replaced in the super
and the queen started oviposition on the FLUVALINATE combs.
More than 1000 cells containing eggs were mapped on the combs of every
colony and seperated from the queen. 10 days later the number of pupae
in these cells were counted.
Regularly the inserts beneath the colonies were checked for dead mites
and dead bees. At the end of the experiment the colonies were killed and
the surviving mites determined.
A group of colonies was kept on FLUVALINATE combs during the summer to
demonstrate the practicability of the treatment. All colonies accepted
the FLUVALINATE comb foundation and started comb building without delay.
All 10 foundations were drawn out to complete combs within a week. The
development of the colonies and the honey production was not different
compared to a group of 5 control colonies (wax foundations without
FLUVALINATE) which were treated equally.
The results are summarized in table 1.

Tab.1 Summer trial
a) Lublin
Start: June 3rd., 1985, end: September 16th.,1985.

Col.No.	n bees	brood survival %	n varroa	rest	success%
P1	35000	81	5581	12	99.78
P2	40183	91	6823	10	99.85
P3	40183	84	4821	16	99.78
P4	32185	94	7330	14	99.81
P5	37301	91	10421	23	99.78
contr.a	39101	82	4912	4231	13.87

b) Oberursel
Start: June 3rd., 1985, end: September 19th., 1985.

Col.No.	n bees	brood survival %	n varroa	rest	success%
01	13900	79	699	00	100.00
02	16400	67	333	02	99.40
03	10700	89	154	01	99.35
04	12100	87	548	01	99.82
05	10500	91	423	01	99.78
contr.b	12400	78	414	393	05.07

OVER WINTER EXPERIMENT.
The critical period in the life cycle of a honey bee colony in Europe is
spring, when the change over of long living winter bees to the summer
bees takes place. This change over has a great impact on the honey
production. If the colony strength stays above a critical level of about
10,000 bees beekeepers can expect a good harvest from cherries, apples
and rape. So we left a group of experimental colonies over winter and
examined them in April or June 1986.

234

OVER WINTER TRIAL.
a) Poland
Start: June 3rd., 1985, End: April 29th., 1986

Col.No.	n bees	n varroa	rest	success %
plw	18000	5421	4	099.93
p2w	18500	4411	0	100
p3w	19000	2036	0	100

b) Oberursel
Start: May 30th., 1985, End: June 26th.,1986

Col.No.	n bees	n varroa	rest	success %
90	27000	299	0	100
67	13000	497	1	99.80
56	23000	281	1	99.64
85	15000	1448	4	99.72
82	17000	423	1	99.76

The manufacturing of bees wax foundations containing 0.05% FLUVALINATE does not require any special equipment. FLUVALINATE does not change the mechanical or optical properties of the foundation.A permanent identification of all frames shall be recommended. This is essential to keep FLUVALINATE combs apart from the normal combs.

No indication of a delay in comb building or any repellent effect was observered in the experiments. The FLUVALINATE foundations were accepted as fast as normal foundations in control colonies.

The success of the mite control was over 99% in all treated colonies. The observed range between 99.35% and 100% is extremly small and indicates the high efficency of the method. There is no other treatment available at the moment which offers comparable results and reliability. Alltogether the results of the field trial indicate that FLUVALINATE foundations offer a chance to eliminate varroa regionally if all colonies are treated simultanously.

The work reported here is still in an experimental stage and there are many questions to be answered before the methods are ready for official registration and general application. But the results demonstrate that the use of carriers or impregnated wax foundations with its prolonged effectiveness has advantages compared to the presently practised treaments. So, this new approach to chemotherapy will bring considerable improvements in controlling varroatosis, the greatest problem of beekeeping in Europe.

LITERATURE

KLEPSCH A., MAUL V., PETERSEN N., KOENIGER N., GÖTZ W., 1983 :
        Feldversuch zur Varroatosebekämpfung mit FOLBEX VA NEU.
        Die Biene 119 (2), 54-57

RITTER W., 1985
> Bekämpfung der Varroatose mit PERIZIN, einem neuen
> systemischen Medikament.
> Apidologie 16(3), 219-220

RUTTNER F., RITTER W., GÖTZ W., 1980:
> Chemotherapie der Varroatose über die Haemolymphe der Biene.
> Allgemeine Deutsche Imkerzeitung 14 (5), 160-165

SEELEY T.D., 1979:
> Queen substance dispersal by messenger workers in honey bee
> colonies.
> Behav Ecol Sociobiol 5, 391-415

WACHENDÖRFER G., FIJALKOWSKI J., KAISER E., SEINSCHE D., SIEBENTRITT J.,
> 1985:
> Labor- und Feldversuche mit der Illertisser Milbenplatte als
> neue Anwendungsform der Ameisensäure im Rahmen der
> Varroatose-Bekämpfung.
> Apidologie 16(3), 291-306

Economical aspects of the 'trapping comb technique' as a new form of bee management

V.Maul

Hessische Landesanstalt für Leistungsprüfungen in der Tierzucht Neu-Ulrichstein, Abteilung für Bienenzucht, Kirchhain, FR Germany

Summary

Procedural variations of the technique of eliminating Varroa mites from the colony during the honey flow season by means of defined combs of worker brood are shortly discussed. Presented data indicate a positive effect of the temporary brood limitation on the honey yield of the running season and, possibly, a slight negative influence on the following year's crop. Only for two of the total of five strictly scheduled management operations, the required working time is higher than in corresponding operations of regular management. The effect on the Varroa population is strong enough to allow the dispension from the otherwise obligatory fall treatment with medicals at least for every second year. Arguments for and against the trapping comb technique as a form of routine bee management are discussed.

1. Introduction

Among the biotechnical methods of Varroa control, the trapping comb technique has turned out to be the most effective one. As originally designed by Ruttner and collaborators (18, 19, 20), the oviposition of the queen is limited over a period of 3 - 4 weeks successively to a few worker combs (trapping combs = Bannwaben), each of which is removed and destroyed shortly before hatching. The technique was primarily designed to serve as a subsidiary measure for the preparation of broodfree colonies which were needed for accurate testing of the efficiency of various chemical treatments. Showing, however, a surprisingly high efficiency of Varroa elimination itself (2, 7, 8, 13, 18, 19, 20), the trapping comb method had to be considered further as a very promising direct approach in Varroa control. A dominant advantage was seen in the chance to minimize the need of any chemical treatment and, in consequence, to minimize the possible risks of residue contamination of honey and wax. Undoubtedly, this aim fits best to an emergency situation, where already during the honey flow season a strong Varroa infestation signalizes the danger of colony losses towards the outrunning season or of severe damage to prospective winter bees (11). In this situation, the possibility of saving the colonies and the full honey crop gives the strongest positive argument. Whether, however, in the sense of a preventive measure, the regular integration of the trapping comb technique into bee management may be profitable over longer periods is still an open question. The present paper tends to show up the positive and negative arguments and to summarize data available up to now which are needed for an evaluation.

2. Timing and procedural variation

The standard technique uses simple excluder frames, fitting to both sides of a regular comb (9). The queen is limited to one empty worker comb four times for seven days (7-day cycle) or three times for nine days (9-day cycle). After the first interval of oviposition, each trapping comb remains in the colony for a second interval to be sealed and is then removed and destroyed. The previously produced brood hatches completely during the first three weeks of treatment. During the fourth to seventh week, no brood will hatch within the colony with the consequence of a more or less marked depression of colony strength later on.

For areas with usual summer honey flow, the procedure should be initiated at or shortly before the point of maximum colony development as signalized by the first expressions of swarming mood. In Central Europe this point is usually reached between end of May and mid of June. The existing brood reserves will then guarantee a full honey crop (9, 14, 15, 20), swarming tendencies will be totally cut down, and sufficient time is left during later summer to build up strong winter colonies. Due to the inevitable colony depression later on, the technique does not fit for areas with late honey flows.

Instead of limiting the queen to single combs, she may be limited by excluder to a super established with one comb and otherwise foundation only, from which subsequently sealed brood is removed (9). Kroos proposed an alternative procedure making queen nuclei from each colony and returning at intervalls open brood from the nuclei to the queenless colonies to serve as Varroa traps (4). For the latter techniques, however, no experimental data are available yet.

3. Effects on the running honey yield and swarming mood

Temporary brood limitation during running honey flow is a traditional means to increase the honey yield (14, 15). The good summer honey flow in 1986 offered good testing conditions for this aspect. A total of 28 colonies in 3 closely neighboured locations was divided according to descendance and colony strenght into two almost equally composed groups of 18 experimentals and 10 controls. The low level of Varroa infestation allowed to place in each location experimental colonies together with control colonies. All colonies had inseminated queens from the institute's Carniolan stock. The treatment ran from June 16th to July 14th (9-day cycle). The honey yield was determined by weighing the honey combs before and after extraction and adding the estimated stores remaining after the last extraction. (1-5 kg).

TABLE I. Comparison of initial colony strength and honey yield from trapping comb treated colonies (n = 18) and controls (n = 10) Treatment: 16.6.-14.7.86, 9-day cycle.

	Experimental $\bar{x}$ (range)	Control $\bar{x}$ (range)
Colony strength 16.6.86		
No. of covered combs	27,2 (21-36)	28,6 (20-38)
No. of brood combs	11,8 (8-18)	13,3 (9-17)
Honey yield 86 (kg)	44,7 (12,8-65,5)	36,9 (16,3-58,9)

The results clearly show an increase in honey yield, caused by the trapping comb treatment (table I). Although some colonies showed initial swarming mood at the starting point, it disappeard completely in both groups without special efforts. This may be due to the used stock with low swarming tendency and the good honey flow as well.

4. Effects on wintering and next year's honey yield

From earlier experiments it is known that brood production rises very sharply after the limitation period and tends to last longer into fall than in regular colonies (12). Despite a slightly lower wintering strength, the experimental group showed a better performance and honey yield during the following season. In these experiments, however, experimentals and controls were placed at different locations so that environmental influences could not be clearly eliminated. The first experiment with both groups in the same locations started in 1985 (10.6.; 9-day cycle). There was a wide range of colony development due to the restrictive wintering and spring conditions before (see table II, colony strength 30.5.). For both groups, equal rates of strong and weak colonies were chosen. The expected summer honey flow was completely missing because of constantly rainy weather.

Under these extremely unfavourable conditions, the data of colony strength at various times and of winter losses showed a clear disadvantage of the experimental group (table II). Nevertheless, there was only a slight and not significant negative balance in honey yield.

TABLE II. Comparison of colony performance and honey yield of the second year from trapping comb treated colonies (n = 21) and controls (n = 12). Treatment: 10.6.-8.7.85, 9-day cycle.

	Experimental $\bar{x}$ (range)	Control $\bar{x}$ (range)
Colony strength 30.5.85		
No. of covered combs	23,2 (10-30)	22,7 (15-32)
No. of brood combs	11,6 (5-17)	11,1 (8-16)
Colony strength 19.9.85		
No. of covered combs	15,1 (7-26)	19,5 (14-25)
No. of brood combs	1,8 (0- 4)	0,8 (0- 2)
Winter losses 85/86 (total)		
Dysentery	1	0
Queenless	2	0
Colony strength 7.5.86		
No. of covered combs	13,6 (5-20)	15,3 (6-18)
No. of brood combs	6,9 (3-12)	8,2 (3-10)
Honey yield 1986 (kg)	38,6 (15,4-62,2)	43,6 (12,8-66,9)

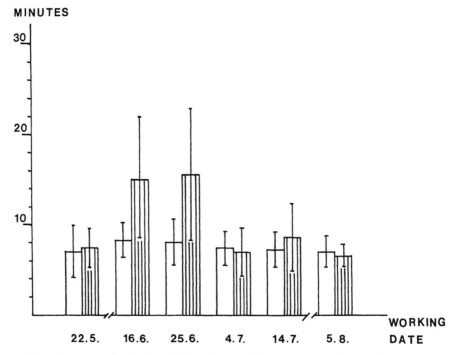

Fig. 1. Comparison of working times for regular management
(light columns, n = 10) and trapping comb treatment
(dark columns, n = 18). Bars indicate standard deviation s.
For further explanation see text.

Starting the same experiment under more favourable conditions of ini-
tual colony strength, however, Staemmler (21) reports on a clear advantage of the experimental group in the following year's honey yield. This
underlines the importance of choosing the adequate starting point.

5. Additional labour
    The major argument of practical beekeepers against this technique is
that the procedure is too much time consuming and bound to strictly fixed
working dates. In a first approach, we have tried to count the actual working time per colony. These data cannot be considered as representative,
however, since under our institute's working conditions each management
operation requires the perceptance of several data for each colony. In
general, a wide range of individual working speed must be expected, depending on personal skill, particularly in finding and handling the queen,
but also depending on the colonies' aggressivity, swarming tendency and
environmental conditions. Nevertheless, the present data may be helpful
in relative comparison for a first orientation (fig. 1).
    Operation 1 (22.5.) is chosen as an example of a regular inspection
with expectedly no difference between the groups. The following operations
of the control group (light columns) are of the same character and always
include a routine inspection on swarm cells. For the experimental group
(dark columns), operation 2 (16.6.) marks the initiation of the treatment.
The queen must be searched and caged to the first trapping comb. At ope-

240

ration 3 (25.6.), the queen should be easily found to be placed on the second trapping comb. Another very thorough inspection of the total brood nest is necessary, however, to find and destroy all swarm cells eventually present. If a virgin has a chance to hatch, the caged queen will be usually killed. Later on, such inspections are no longer necessary, but still may be required under regular management.

Operation 4 (4.7.) requires the installation of the third trapping comb and the removal of the first one. At operation 5 (14.7.), the second trapping comb is removed, and the queen is set free again or, at this point, very easily, replaced by a young one. Since no brood is present except on trapping combs, an optimal chance is given to reassemble the combs of the brood chamber for the subsequent brood period.

The removal of the third trapping comb, a very short procedure only, was not registered accidentally and cannot be shown, therefore. Operation 6 (8.8.) represents again a routine inspection of the later season, preceding the feeding period.

Apparently, only the two initial steps of the procedure require more working time than usual. The variation is high in both cases, as can be expected. An advantage may arise later on from saving efforts in the control of swarming. In 1986, however, we had no chance to show this because of lacking swarm tendency in all of the controls.

6. Effects on the Varroa population

The efficiency of Varroa elimination could be shown to range roughly between 80 % and 95 % (2, 7, 8, 13). During the following brood cycles, lasting over at least two months, the residual mites and probably reinvading mites from neigbouring untreated apiaries have a good chance to propagate.

Preliminary estimations from a small private apiary, run with trapping comb technique in a heavily infested area, indicate that at the end of the season the infestation level of the starting point may be reached again or even surpassed (results of chemical fall treatment). In our own experimental apiary, also in a heavily infested area, we try since 1985 to maintain the colonies without additional fall treatment. The only means to relatively estimate the changes of infestation level is the consequent registration of naturally dying mites. It must be kept in mind, however, that besides the weak correlation between mite mortality and population size of Varroa (1, 5, 6, 10, 11, 16, 17) there are seasonal effects on this character (increase of mite mortality in fall). The data are summarized in figure 2 as monthly average of dead mites/day. Superposed bars indicate the active time of the three trapping combs.

Registration started in May 1985 at a still moderate level of Varroa infestation. The data indicate a slight reduction of the Varroa population by the treatment. The following peak (September - October) combines the effects of the new increase of the Varroa population and the seasonal increase in mite mortality.

In the rising season of 1986 (second row of columns), the Varroa population is makedly stronger. From washing all trapping combs from 5 colonies, it can be estimated to range between 2000 and 7000 mites/colony at the maximum. This infestation level must be considered as an emergency situation already. The treatment appears to be very effective, cutting the Varroa population down almost to the level of the corresponding period the year before. Beyond September no data are available yet. It can be presumed that the winter population of mites will be about twice as strong as the year before. Nevertheless, an additional fall treatment will be omitted again in 1986. It will become necessary very probably, however, in 1987.

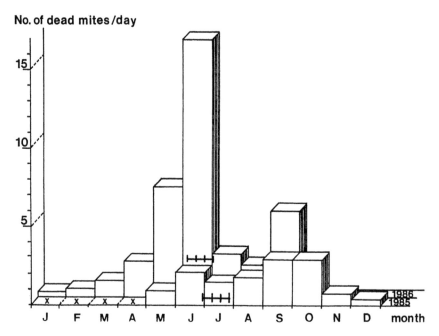

No. of dead mites /day

Fig. 2. Influence of trapping comb treatment on the Varroa population.
Natural mite mortality, monthly calculated as average no. of dead
mites/day, is used as an indicator (n = 11). Superposed bars
indicate the times of active trapping combs.

So far, no sign of harm could be detected in any of the colonies. The
honey yield 1986 amounted to an average of 47,7 kg/colony.

7. Discussion

Future bee management inevitably must include some kind of regular
Varroa control. On these premises, an objective evaluation of a routinely
integration of the trapping comb technique into bee management should take
into consideration also the labour and cost of a regular fall treatment.
This has so far been omitted in this study.

The following positive arguments can be listed unquestionably:
a) Efficient Varroa control during the season, even at a high infestation level.
b) Efficient control of swarming.
c) Probable increase of honey yield from the running season.
d) In case of an emergency, reduced input in chemical treatment and, in
   consequence, reduced risk of honey contamination by residues.

All possible negative arguments, on the other hand, appear to have a
relatively variable weight according to the individual situation.
e) The surplus of labour required at the two initial working steps may
   be reduced by a skilled working routine and, on the other hand, may be
   partly balanced by the reduction of labour for swarming control. Eventually, also the labour of a regular fall treatment may be reduced.
f) For all colonies of an apiary, strict and synchronous timing is
   obligatory. A professional beekeeper will prefer the 9-day cycle,
   whereas the 7-day cycle should be more comfortable for the parttime
   beekeeper working on weekends. Depending on the available time during
   the critical period, there is of course a certain limit of colonies

to be managed. This, however, cannot be generalized, but must be found out individually.

g) The effects on the following year's honey yield may be variable. But again, a skilled management may help to avoid negative effects, for example by choosing the optimal starting point and equalizing the colonies of each apiary before this date.

Taking all arguments together, the trapping comb technique indeed appears to be a very promising way of future bee management. As long as state control of Varroa is maintained, however, the possible dispension from a chemical fall treatment will depend on an individual permission by the authorities. If the beekeeper is responsible himself, he will follow economical arguments only. For both situations, a practical system for a reliable evaluation of the actual infestation level is of essential importance.

REFERENCES

1. FUCHS, S. and KOENIGER, N. (1984). Rechnen oder Raten - das Dilemma bei der Abschätzung des Varroabefalls. Allgemeine Deutsche Imkerzeitung 18, 294-296

2. KLEPSCH, A. and MAUL, V. (1983). Neue Versuche zur Wirksamkeit des Bannwabenverfahrens. Internatonaler Bienenzüchterkongreß der Apimondia Bukarest 1983, 258-265

3. KOENIGER, N. and SCHULZ, A. (1980). Versuche zur biologischen Therapie der Varroatose durch eine Kontrolle der frischgeschlüpften Bienen. Apidologie 11, 105-112

4. KROOS, H. (1986). Die Wildeshauser Betriebsweise. Private publication (Haus am Kiek, 2832 Rüssen)

5. LIEBIG, G., SCHLIPF, U., FREMUTH, W. and LUDWIG, W. (1984). Ergebnisse der Untersuchungen über die Befallsentwicklung der Varroa-Milbe in Stuttgart-Hohenheim 1983. Allgemeine Deutsche Imkerzeitung 18, 185-190

6. LIEBIG, G. (1985). Ergebnisse des Feldversuchs 1984. Allgemeine Deutsche Imkerzeitung 19, 211-212

7. MAUL, V. (1982). Kurzbericht über Versuche zur Varroatosebekämpfung mit Japanischem Heilpflanzenöl. Die Biene 118, 388-390

8. MAUL, V. (1983). Varroa-Elimination mittels Brutbeschränkung auf Bannwaben - neue Ergebnisse zur Wirksamkeit des Verfahrens (Short communication). Apidologie 14, 260

9. MAUL, V. (1983). Empfehlungen zur Methodik der Varroa-Elimination mittels Bannwaben aus Arbeiterbrut. Allgemeine Deutsche Imkerzeitung 17, 179-194

10. MAUL, V. (1984). Abschätzen des Varroabefalls über den spontanen Milbenabfall (short communication). Apidologie 15, 243-244

11. MAUL, V. (1984). Empfehlungen zu Vorsorgemaßnahmen gegen Varroaschäden im Sommer 1984. Die Biene 120, 249-253 + 298-300

12. MAUL, V., SCHNEIDER, M. and ECKERT, D. (1984). Biologische Bekämpfung der Varroatose mittels Bannwabenverfahren. Hessische Landesanstalt für Leistungsprüfungen in der Tierzucht Neu-Ulrichstein, Jahresbericht 1984, 62-63

13. MAUL, V. and KLEPSCH, A. (1985).Biologische Bekämpfung der Varroatose. Hessische Landesanstalt für Leistungsprüfungen in der Tierzucht Neu-Ulrichstein, Jahresbericht 1985, 72-76

14. PREUSS, E. (1919). Meine Bienenzucht - Betriebsweise und ihre Erfolge. Potsdam 1899 (1. Aufl.), 1900 (2. Aufl.), Leipzig 1919 (3. Aufl.)

15. PREUSS, CH. (1920). Preuss'sche Imkerschule. Verlag T. Fischer, Leipzig, Berlin, Freiburg, 1920

16. RADEMACHER, E. (1985). Untersuchungen zu einem Prognose-Modellversuch anhand des natürlichen Totenfalles von Varroa Jacobsoni (short communication). Apidologie 16, 215

17. RADEMACHER, E. (1985). Ist eine Befallsprognose aus dem natürlichen Totenfall von Varroa Jacobsoni möglich? Apidologie 16, 395-405

18. RUTTNER, F. and KOENIGER, N. (1979). Versuche zur Eliminierung der Varroa-Milben mit biologischen Methoden. Internationaler Bienenzüchterkongreß der Apimondia Athen 1979, 400-402

19. RUTTNER, F. and KOENIGER, N. (1980). Eine biologische Methode zur Eliminierung der Varroa-Milben aus Bienenvölkern. Allgemeine Deutsche Imkerzeitung 14, 11-12

20. RUTTNER, F. KOENIGER, N. and RITTER, W. (1980). Brutstop und Brutentnahme. Allgemeine Deutsche Imkerzeitung 14, 159-160

21. STAEMMLER, G. (1986). Das Bannwabenverfahren, unter schleswigholsteinischen Witterungsbedingungen getestet. Die neue Bienenzucht 13, 11-14

Conclusions

Conclusions and recommendations on general discussion

N.Koeniger

Institut für Bienenkunde (Polytechnische Gesellschaft), Fachbereich Biologie, J.W.Goethe Universität Frankfurt am Main, Oberursel, FR Germany

Within the short space of five years since the inauguration of the EC Varroatosis research programme, the Varroa jacobsoni mite has spread westward to infest a large number of EC-Countries.

With the exception of Great Britain, Ireland, and Portugal, it is now well established within the community, positive justification for the timely decision upon the part of the CEC to support researches to combat the problem.

After a wide-ranging discussion, the participants agreed on the following.

The problems influencing the progress of the research programme are many. After more than a century of study our understanding of the life and ways of the honey-bee is still incomplete. Yet, we have suddenly been faced with investigating the biology and behaviour of a mite new to both the bee and to us. To add to the difficulties it is necessary to expend a considerable amount of the total available resources on producing urgent palliative methods to keep the pest under control.

This, in turn, introduces difficulties of finding an active ingredient poisonous only to one of a pair of organisms living as commensals.

Moreover, since honey is and always will be considered a pure natural food, in this time of awareness over environmental pollution we need to ensure that control strategies are harmless to the consumers of the honey.

Finally, our task is hampered by the limited resources available.

At the last bid for monetary support, a call for research projects yielded 28 responses from EC-research institutions. Available funds could only support eight of these projects, but at about half the budget level requested. This has severely limited the over-all scope of the programme, resulting in the contraction of important project areas.

Nevertheless progress is recorded as the meeting has shown.

A good range of proprietary products, involving a number of different active ingredients, are now commercially available or hopefully will shortly become available to beekeepers. Their availability can be directly attributed to the screening and development studies carried out within the Varroatosis research programme.

On the biological front we are developing an understanding of the reproductive capabilities of Varroa. As well as some of the mite's behavioural factors and biochemical changes occuring during the development of the host, which relate to successful reproduction.

The input on the study of virus and bacterial infection associated with Varroa infestation is limited, even so it has been possible to establish that there is a relationship between the presence of acute paralysis virus and the level of Varroa infestation.

Finally, following earlier recommendations a study has at last been initiated to develop an artificial rearing technique to produce populations for international standardised testing procedures. Initial results are promising in that feeding by all stages through a membrane on an artificial diet has been achieved.

It is considered that within the restraints of present resources the progress of the five-year research programme is very satisfactory.

The priority areas referred to above and in previous reports remain unchanged. However, as a result of this meeting it has become evident that studies need to be undertaken to monitor resistence build-up and, if possible, to countermand the effect.

We recommend a small group of experts should meet to consider how to develop a standard discriminant dosage test.

It also become obvious from the discussons within the meeting that factors affecting the speed of diffusion of the infestation of a country and the rate of build-up of populations need investigation, particularly relative to the importance of beekeeping practices and climatic conditions.

The bi-annual meeting of EC-delegates has revealed the importance of personal contact between scientists across the breadth of the programme. It is also apparent that some mechanism should be available for smaller groups of experts in a particular field of study to meet for short discussions at more regular intervals. The severe monetary restraints within current project budgets makes this extremely difficult.

We ask the CEC to give consideration to supporting such "exchange of ideas" meetings essential to the success of projects from which results are urgently required.

It is considered useful to have next experts' group meeting in October 1988. The Italian delegate Prof. F. Frilli has formally invited the next meeting to take place at the University of Udine, Italy, and participants request the CEC to realize it.

. . . .

A study tour was organized in the field and to the beekeeping Institutes in Oberursel and in Kirchhain.

The demonstration in Oberursel were centered on screening and testing of new substances. The maintainance of colonies with a high <u>Varroa</u> infestation in flight rooms was demonstrated. Further the participants saw the filling and the handling of the disposable screening cages. Another demonstration was offered on a new method to determine the exact number of mites from a bee colony. Artificial insemination of honey-bee queens and experiments on genetics gave an impression of the 'Non-Varroa' research work of the Institute which belongs to the faculty of biology of the University of Frankfurt.

On the way to Kirchhain a field experiment was visited at the Saalburg mountain. Plastic carriers impregnated with Bayvarol, a new pyrethroid, were placed between the combs in the center of brood nests. This position had a higher efficiency compared to carriers above and beneath the combs or at the hive entrance.

At the Kirchhain Institite Dr. V. MAUL and A. KLEPSCH presented the results of a long term study of the distribution of <u>V. jacobsoni</u> in Hessen. The visitors were very much impressed by the efficient cooperation between the Institute and the beekeepers. The valuable data showed significant differences in population development among the regions.

Another highlight of the visit was the demonstration of the trapping comb technique. This control method was tested for more than 3 years at Kirchhain. The efficiency was over 90%.

A 'sight seeing' tour through the Institute, from the honey analysis laboratory, through the artificial insemination room, and the exhibition room of local old and new hives and equipment, concluded the stay.

The German Beekeeper's Association (DIB), to honour the distinguished visitors, kindly offered a charming reception at the castle of Rauschholz-hausen.

List of participants

Belgium:

CASTEELS Peter R.
Laboratory for Zoophysiology - State University Ghent
K.L. Ledeganckstrasse, 35
9000 GHENT

DE GREEF Myriam
Rijksstation voor Nematologie en Entomologie
Burg. van Gansberghelaan, 96
9220 MERELBEKE

DE WAEL Lutgarde
Rijksstation voor Nematologie en Entomologie
Burg. van Gansberghelaan, 96
9220 MERELBEKE

HAVAUX Jean-Claude
SPB, Sanders-Probel Biotechnology
H. Wafelaerts, 47-51
1060 BRUXELLES

HERWIG Ramon
Rijksstation voor Nematologie en Entomologie
Burg. van Gansberghelaan, 96
9220 MERELBEKE

JACOBS Frans J.
Laboratory for Zoophysiology - State University Ghent
K.L. Ledeganckstrasse, 35
9000 GHENT

VAN LAERE Octaaf
Rijksstation voor Nematologie en Entomologie
Burg. van Gansberghelaan, 96
9220 MERELBEKE

VAN STEENKISTE Danny
Laboratory for Zoophysiology - State University Ghent
K.L. Ledeganckstrasse, 35
9000 GHENT

Denmark:

HANSEN Henrik
Statens Bisygdomsnievn
Virumgaard-Kongevejen, 83
2800 LYNGBY

Federal Republic of Germany:

BERG Stefan
Egestrasse, 27
6000 FRANKFURT

BUECHLER Ralph
Institut fuer Landwirtschaftliche Zoologie und Bienenkunde
Melbweg, 42
5300 BONN

FUCHS Stefan
Institut fuer Bienenkunde der Universitat Frankfurt
Karl-von Frisch-Weg, 2
6370 OBERURSEL

GNADINGER Fridolin
Deutscher Imkerbund
Am Bildstock, 16
7768 STOCKACH

KIMMICH Karlheinz
Universitat Hohenheim Landesanstalt fuer Bienenkunde
August von Hartmannstrasse
7000 STUTTGART

KLEPSCH Andreas
Hessiche Landesanstalt fuer Leistungsprufungen in der Tierzucht
Neu-Ulrichstein, Abteilung fuer Bienenzucht
Erlenstrasse, 9
3575 KIRCHHAIN

KOENIGER Nikolaus
Institut fuer Bienenkunde der Universitat Frankfurt
Karl-von Frisch-Weg, 2
6370 OBERURSEL

KRIEGER Klemens
Zentrum Landwirtschaft
Alfred Nobel Strasse
5090 LEVERLENSEN

MAUL Volprecht
Hessiche Landesanstalt fuer Leistungsprufungen in der Tierzucht
Neu-Ulrichstein, Abteilung fuer Bienenzucht
Erlenstrasse, 9
3575 KIRCHHAIN

MAUTZ Dietrich
Bayer, Landsanstalt fuer Bienenzucht
Burg. Bergstrasse, 70
8520 ERLANGEN

OTTEN Christoph
Institut fuer Bienenkunde der Universitat Frankfurt
Karl-von Frisch-Weg, 2
6370 OBERURSEL

PITTLER Hermann
Bundesministerium fuer Ernahrung, Landwirtschaft und Forsten
Rochusstrasse, 1
5300 BONN

RADEMACHER Eva
FUB, Institut fuer Allgemeine Zoologie
Konigin-Luise-Strasse, 1-3
BERLIN

RITTER Wolfgang
Tierhygienisches Institut
Am Moosweiher, 2
7800 FREIBURG

SAKOFSKI Fritz-Michael
Institut fuer Bienenkunde der Universitat Frankfurt
Karl-von Frisch-Weg, 2
6370 OBERURSEL

SCHIEFERSTEIN ERICH
Dutscher Imkerbund
Am Ritterkelle, 25
6368 BAD VILBEL

SCHNEIDER Petra
Institut fuer Landwirtschaftliche Zoologie und Bienenkunde
Melbweg, 42
5300 BONN

France:

ARNOLD Gerard
Institut National de la Recherche Agronomique - CNRS
Laboratoire de Neurobiologie Comparée des Invertébrés
Rue de la Guyonnerie, 1
91440 BURES-SUR-YVETTE

BORNECK Raymond
Institut Technique de l'Apiculture - ITAPI
Rue de la Guyonnerie, 1
91440 BURES-SUR-YVETTE

LECLERCQ Pascal
France-Miel
B.P. 5
39330 MOUCHARD

LE CONTE Yves
Institut National de la Recherche Agronomique - CNRS
Laboratoire de Neurobiologie Comparée des Invertébrés
Rue de la Guyonnerie, 1
91440 BURES-SUR-YVETTE

Great Britain:

BALL Brenda V.
Rothamsted Experimental Station
HARPENDEN AL5 2JQ

GRIFFITHS Donald A.
Ministry of Agriculture, Fischeries and Food - ADAS
Slough Laboratory
London Road
SLOUGH SL3 7HJ

HUGHES Robert G.
7, Chertsey Road
WOKING

Greece:

IFANTIDIS Michael
Laboratory of Apiculture - School of Agriculture - University
D. Mylona, 6
54636 THESSALONIKI

SANTAS Loukas
Laboratory of Sericulture and Apiculture
College of Agricultural Sciences - University
Votanicos
ATHENS

SOULIOTIS Kostas
"Benaki" Phytopathological Institute
Ekalis, 2
14561 KIPHISSIA

TSELLIOS Demetrios
Agricultural Research Station of Halkidiki
63200 NEA MOUDANIA

Ireland:

HUME Hubert Desmond
Department of Agriculture - Agriculture House
Kildare Street
DUBLIN

Italy:

ACCORTI Marco
Istituto Sperimentale per la Zoologia Agraria
Via di Lanciola, 15
50125 CASCINE DEL RICCIO

BRUCE William A.
Istituto di Difesa delle Piante - Università
P.le M. Kolbe, 4
33100 UDINE

CHIESA Fiorella
Istituto di Difesa delle Piante - Università
P.le M. Kolbe, 4
33100 UDINE

FRILLI Franco
Rettorato - Università degli Studi
Via Antonini, 8
33100 UDINE

GALLO Carmelo
Istituto di Zootecnica - Università
Viale delle Scienze, 2
90128 PALERMO

MARCHETTI Stefano
Istituto di Produzione Vegetale - Università
P.le M. Kolbe, 4
33100 UDINE

MILANI Norberto
Istituto di Difesa delle Piante - Università
P.le M. Kolbe, 4
33100 UDINE

TACCHEO BARBINA Maria
Centro Regionale Sperimentazione Agraria
Via Sabbatini, 5
33050 POZZUOLO DEL FRIULI

Netherlands:

BEETSMA Joop
Department of Entomology and Virology - Landbouwhogeschool
Binnenhaven, 7 - P.O. BOX 8031
6700 WAGENINGEN

DE RUIJTER Arie
Research Centre for Insect Pollination and Beekeeping "Ambrosiushoeve"
Tilburgseweg, 32
5081 HILVARENBEEK

VAN HEEST-VERLOOP Anneke
Beedisease Committee
Op ten Noortlaan, 3
6711 EDE

WIEGERS Frans P.
Department of Entomology - Agricultural University
Binnenhaven, 7 - P.O. BOX 8031
6700 WAGENINGEN

Spain:

GOMEZ PAJUELO Antonio
Centro Experimental Agrícola y Ganadero
Diputacion de Cadiz
Apartado de correos, 103
JEREZ DE LA FRONTERA

Yugoslavia:

KULINCEVIC Jovan M.
Pcelarski Kombinat
Jovana Rajica, 5
11000 BEOGRAD

C.E.C.:

CAVALLORO Raffaele
Commission of the European Communities
"Integrated Plant Protection" Programme
Joint Research Centre
21020 ISPRA (ITALY)

List of authors

Accorti, M. 209
Arnold, G. 41

Ball, B.V. 95
Barbina Taccheo, M. 131
Beetsma, J. 57
Borneck, R. 223
Bruce, W.A. 125

Casteels, P.R. 105
Cavalloro, R. 3, 11
Chiesa, F. 113, 125
Chmielewski, M. 183, 185, 231

D'Agaro, M. 131, 145
De Greef, M. 15
De Paoli, M. 131
De Ruijter, A. 63
De Wael, L. 15, 195

Frilli, F. 29, 145
Fuchs, S. 69, 73, 77

Gallo, C. 173
Genduso, P. 173
Gliński, Z. 183, 185
Gnädinger, F. 19
Gomez Pajuelo, A. 35
Griffiths, D.A. 11

Hume, H.D. 27

Ifantidis, M.D. 195, 203

Jacobs, F.J. 51, 105

Klepsch, A. 213
Koeniger, N. 81, 231, 247
Kulincevic, J.M. 177

Le Conte, Y. 41

Marchetti, S. 131, 145
Maul, V. 237
Milani, N. 113
Molins Marín, J.L. 35
Müller, K. 77

Otten, C. 69

Pappas, N. 85, 203
Perez García, F. 35
Prota, R. 31, 161

Ramon, H. 195
Ritter, W. 157
Roth, H. 157

Sakofski, F. 81
Santas, L.A. 23, 187
Schieferstein, E. 7

Thrasyvoulou, A. 85, 203
Tomazin, F. 177

Vandergeynst, F. 51
Van Esch, J. 57
Van Laere, O. 15, 195
Van Steenkiste, D. 105

Wiegers, F.P. 99

Milton Keynes UK
Ingram Content Group UK Ltd.
UKHW051933141024
449569UK00027B/1473